Felix Publishing 2017
www.felixpublishing.com.au
email: info@felixpublishing.com
Print copies available from publisher.

EXPLORATION SCIENCE
Part of the Series **ADVENTURES IN EARTH SCIENCE**
Other books in the series include:
> Riches from the Earth (Minerals, Mining & Energy)
> Changing the Surface (Erosion and Landscapes)
> Rocks – Building the Earth
> Fossils – Life in the Rocks
> A Dangerous Planet (Earth Hazards)
> Through Sea and Sky
> Beyond Planet Earth (Astronomy)

2016 digital book release
ISBN: 978-0-9946432-7-8
Print Edition
ISBN: 978-0-9946432-8-5
Author: Dr P.T.Scott
All illustrations, photographs and videos by the author unless stated
Cover photo: Obtaining a sample of fresh lava, Kilauea Volcano, Hawaii
(Photo: USGS), Design based on that of AJS Creative, Brisbane.

Registration:
Thorpe-Bowker +61 3 8517 8342
email: bowkerlink@thorpe.com.au

FELIX
PUBLISHING

EXPLORATION

SCIENCE

Dr. Peter T. Scott

FELIX
PUBLISHING

First released 2016

To my Grandchildren who are
yet to find their own adventures.

About the Author

Dr. Peter Scott is an award-winning teacher of Earth Science of over forty years' experience in both Secondary and Tertiary Education. He holds a Bachelor's Degree, two Master's Degrees and a Doctorate including many years on his own research in locating and correlating coal measures. Apart from his own exploration during his own research, he has also travelled extensively and has visited many places of interest to Earth Sciences including Antarctica, the Andes, the Amazon, North Africa, volcanic islands of the Pacific and Asia, California northern Europe, Hawaii, and remote Australia.

Dr. Scott, Paradise Bay, Antarctica 2011

Table of Contents

Chapter 1: Geology and its Methods

1.1 Introduction

Earth science is a composite study embracing the sciences of **geology, meteorology, oceanography** and **astronomy.** Earth science attempts to reconstruct the past events of the Earth in order to understand present conditions and problems. From scientific evidence and its evaluation, predictions can be made concerning the nature, dangers and possible use of the Earth's environment. With recent advances in planetary geology, this approach can also be applied to the other planets of our solar system.

These sciences are always developing with old ideas being questioned and new ones emerging with new evidence being discovered. Often the study areas are large and sometimes remote. New technologies coupled with thorough observation and clear thinking lead to new theories. Geology, like the other earth sciences, has always been an exploration-based discipline, with the geologists spending a considerable part of their time away from the laboratory and in the field.

So vast is the study of the Earth, that over the relatively brief, formal development of geology, it has been necessary to sub-divide it into several branches. In some areas of thinking, oceanography is sometimes included within the science of geology and meteorology often is excluded as an independent science. Oceanography and meteorology are explained in THROUGH SEA AND SKY (another book in this series). Geology in turn may be further sub-divided into specialist sciences including:

GEOLOGY		
RESOURCE GEOLOGY	PHYSICAL GEOLOGY	HISTORICAL GEOLOGY
1. MINERALOGY (minerals)	1.TECTONICS (deformations)	1. STRATIGRAPHY (layers of rocks)
2.PETROLOGY (rocks) a.Igneous Petrology b. Sedimentary Petrology c. Metamorphic Petrology	2.VULCANOLOGY (volcanoes) 3. SEISMOLOGY (earthquakes) 4.OCEANOGRAPHY (ocean & sea floor)	2. PALAEONTOLOGY (fossils) a. Palaeobotany b. Invertebrate Palaeontology c. Vertebrate Palaeontology
3. ECONOMIC GEOLOGY (mining, engineering etc.) 4. GEOCHEMISTRY (Chemistry & reactions)	5. GEOPHYSICS (forces in the Earth) 6.ENGINEERING GEOLOGY (construction)	

Table 1.1: Some of the branches of geology

Integrated with these earth sciences are those of other disciplines. For example, a palaeontologist concerned with ancient lifeforms would also have a good knowledge of biology. A geophysicist concerned about the forces within the Earth would also have a considerable knowledge of physics. In addition, to having a good understanding of the science, one must also have a good understanding of the general laws of the total environment and how to survive in it, navigate through it and then represent it as data or on maps.

1.2 The Scientific Method

Earth scientists usually begin their investigations by making up a **hypothesis** - a basic idea about the possible outcome of their study. Using experiments and collected or measured data, from which they can make inferences which are intelligent decisions about what the data means. These ideas may lead to the proof or not, of the hypothesis. Once a hypothesis has been thoroughly tested, it may be up-graded to a **theory** - an idea which seems to be true so far.

Hypotheses and theories are constantly being tested as new experiments and observations are made. If the evidence from these new studies always seems to go against the current ideas then the old hypothesis or theory may have to be changed. If a well-established theory proves to be correct under all conditions of new experimentation and observation, then it may become a scientific **law.** Even then, science changes with new evidence and even well-established laws will be changed. Our understanding is always changing with new evidence which is valid and truthful.

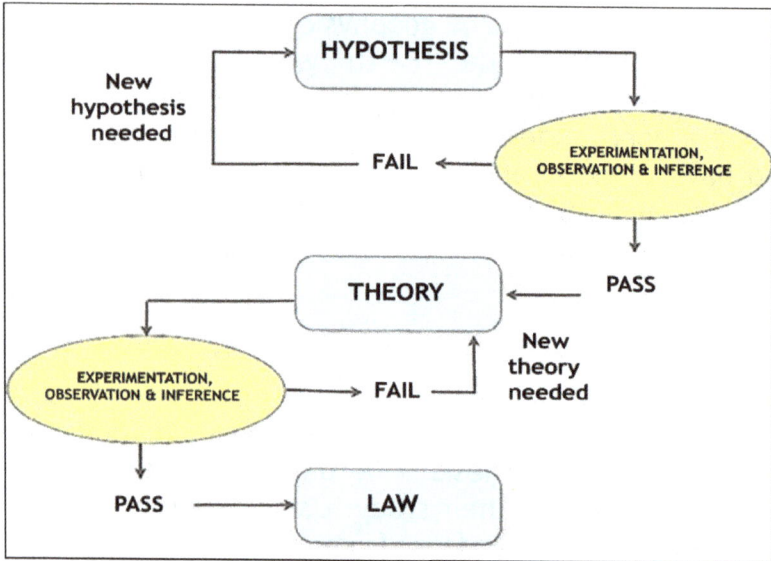

Figure 1.1: A diagram showing the progression of scientific investigation

For example, at the beginning of the nineteenth century, some European geologists thought that rocks which contained large, inter-locked crystals such as granite, had been formed as the waters of the great Biblical flood evaporated, just as salt is left when seawater evaporates. Their original hypotheses eventually came together as the **Neptunist Theory**, named after the Roman god of the sea. Other geologists thought that these rocks were formed from molten rock material which came from deep below the Earth's surface and then cooled. For this reason, their ideas developed into the **Plutonist Theory**, named after the

Roman god of the underworld. After considerable experiment and observation throughout the nineteenth century, geologists established the fact or law that these crystalline rocks did, indeed, come from molten rock cooling at great depth.

The main observations are often made in the field, i.e. outside of the laboratory and in the geological area of study which can then be supported by further studies back in the laboratory. Once all of the data has been studied, hypotheses may be made about the research questions, and then the full study is written up as a report or a scientific paper in a reputable scientific journal.

The experiment method follows a similar, logical format:

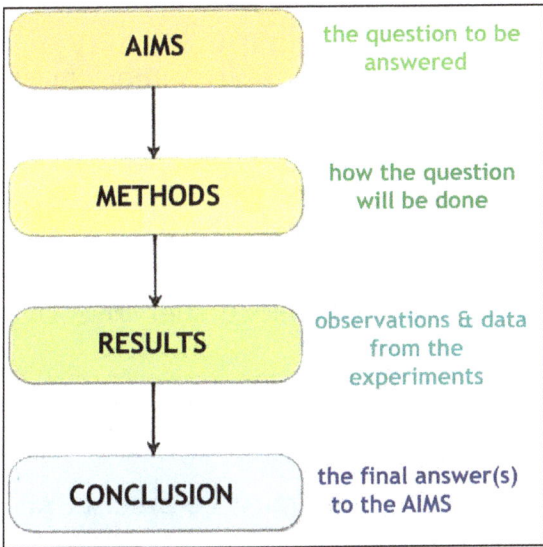

Figure 1.2: Diagram showing the progression of the experimental method

- The Aim sets out precisely what the researcher wants to know. It may be a simple statement such as "To map Colca Canyon", or it may be a series of questions such as "What is the length of Colca Canyon?", "How was it formed?" and "What are the ages of its basement rocks?"

- The Method is the exact procedure which is to be used in the experiment. This could be a simple test such as testing minerals for their hardness, or it may be a complex system involving series of expeditions and laboratory activities. In any case, the method must be detailed enough so that others can later reproduce it exactly.

- The Results include all the data which has been gathered for the total experiment. Field collections, therefore must be very accurately labeled and stored and all numerical data and observations recorded and

- The Conclusion contains the final answers to the Aim. Usually it will also include all of the additional questions which may come from the study and which will need further investigation.

1.3 Instruments

Data gathering in the field must be precise. All readings must be carefully tabled and specimens wrapped and stored along with a thorough record of the:

- Locality

- Geographic orientation, if the rocks or structure is layered or tabular

- Geological significance such as the adjacent rocks, layers, fossils and structures

- Probable classification of the rocks and

- Mineralogy.

To obtain data and collect specimen in the field a variety of scientific instruments are used:

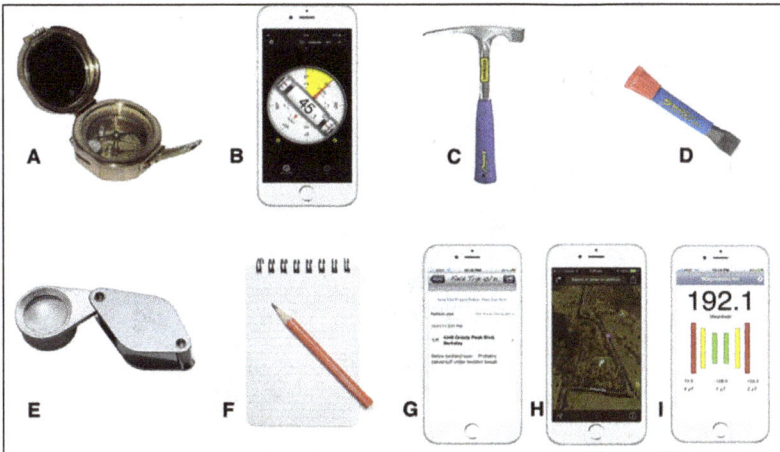

Figure 1.3: Some of the instruments used by a Geologists in the Field including: A. Brunton Compass; B. Smart Phone with Compass App. C. Geological Hammer; D. Rock Chisel; E. Hand Lens; F. Notebook & Pencil; G. Personal Data App.; H. GPS App. I. Data Log App.

Online Video: 1.1: The safe use of a geological hammer
https://www.youtube.com/watch?v=VjQda-q1v1E

The geologist frequently uses more sophisticated devices to collect data. These may include:

- Brunton Pocket Transits (Brunton Compasses) – which measure both magnetic bearings and elevation angles. They are used to measure the angle of tilted linear features, such as rock bedding and faults, as well as the compass direction (or bearing) along which these features run. These are called the **angle of dip** and the **strike** respectively. The angle of dip is measured with an inclinometer spirit level in the compass and the strike is measured using the compass folded out flat.

Figure 1.4: The Brunton Pocket Transit

- Magnetometers- measure small variations of Earth magnetism in different rocks and can be used to locate specific rock types and structures.

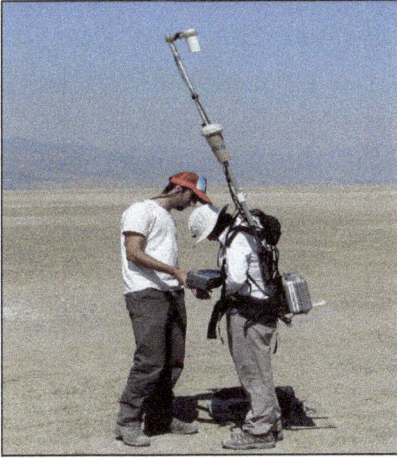

Figure 1.5: A magnetometer being used in a ground survey (Photo: USGS)

- Gravimeters · measure small variations in the Earth's gravitational field due to rocks. They are used in a similar way to the magnetometer and can be used to determine size and density of rock bodies below the surface.

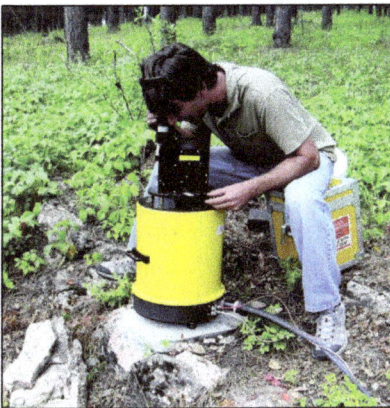

Figure 1.6: Using a gravimeter (Photo: Karl Koth, USGS)

- Electrodes - are used with different methods to measure natural or man-made electrical fields, within the upper crust and there for the nature and type of rock.

- Radiation counters - are used to measure natural and man-made radioactivity for locating radioactive minerals and estimations of sub-surface rock density.

- Drilling equipment · both for land and sea-floor exploration, is used for rock sampling and oil and gas exploration.

Figure 1.7: Research drilling rig (Photo: USGS)

- Geothermal equipment · temperature measuring devices such as thermometers and electronic thermistors for land and marine use. They measure heat flow and for corrections in other devices.

- Survey and mapping equipment - for geological mapping and obtaining precise locations.

- Seismic devices - Seismometers are used for natural and man-made earthquake detection. These are explained in the companion book A DANGEROUS PLANET.

Figure 1.8: A small field seismometer

- Imagery equipment – such as cameras onboard satellites, using natural light, X-rays, infra-red and ultra violet wavelengths. Especially used in the production of geological maps and for the location of ores, rocks and sub-surface structures and the

- Petrological Microscope - used in rock and mineral studies. The specimen is cemented to a glass slide and ground down to about 0.03mm becoming transparent. The microscope functions like a biological microscope except that it uses polarised light and has a rotating stage. This is one of the prime instruments of the geological laboratory. A view through the microscope shows the different minerals as seen in **cross-polarised** light, i.e. the slide is sandwiched between two Polaroid filters which are orientated at 90 degrees to each other.

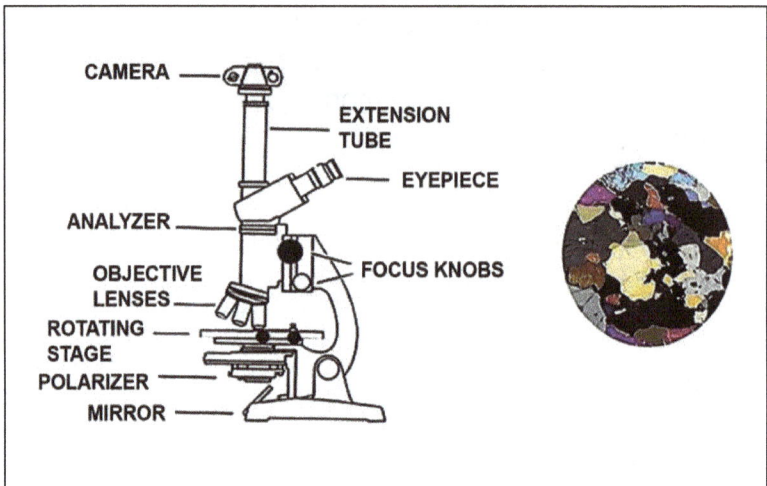

Figure 1.9: Diagram of a petrological microscope (left) and a view of a granite rock thin section using polarised filters (cross-polars) and x 25 magnification

Online Video: 1.2: A demonstration using polarized filters.
Go to https://www.youtube.com/watch?v=SeQzTVdsvrg

Chapter 2: Becoming a Geologist

2.1 Introduction

Unlike many formal sciences, geology has plenty of scope, both as an interesting and profitable profession, but also as a part-time hobby for the average person. Anyone can study the Earth around them, its minerals and natural processes, without any formal training, and yet contribute to their own lifestyle as well as to discipline of science.

Figure 2.1: A volcanologist gathering fresh lava from Kilauea volcano, Hawaii (Photo: USGS)

Geology for both the amateur and the professional often becomes a passion and can be an exciting lifestyle. Sometimes it involves periods of office work and a little hardship in the field, but it is always interesting, especially when field data begins to come together to give a bigger picture of what going on in the Earth. There are many informal study courses at community colleges and through the internet. This book was also written for anyone interested in the subject as well as the student Earth Scientist.

Figure 2.2: Student geologists at a coal mine, Ipswich, Queensland, Australia

Formally, students must be dedicated and interested in the general or specific aspects of geology to become a professional in its many disciplines. Usually a good grounding at secondary school will give a good start, especially in high school Earth Science programs. Chemistry and mathematics are also useful support subjects, but with strong motivation, one can do bridging courses to overcome any personal deficiencies in these subjects. The most important goal is to gain entrance into a college or university or similar schools, such as a School of Mines. Even without prior training, a student can often transfer at a later date into a formal geology program provided they can convince the faculty that they are able.

University courses often take three or four years of full-time study but some offer part-time and distance education courses by correspondence or via the internet. However, such programs also require some additional field work and laboratory programs in support of the theory. Formal studies usually lead to a Bachelor's Degree, but more able students can continue and complete further studies with additional independent research to obtain an Honour's Degree. Further specific study and research can lead to a Master's Degree or for more difficult research, a Doctorate. Higher research degrees usually provide better job prospects, but there are many other factors in taking up employment in geology. These include the availability of employment, the location of research areas and the personal life of the individual which may limit employment choices. Many students who are not prepared to travel or do extensive field work may prefer office or laboratory positions working for government agencies and private enterprise.

2.2 Code of Ethics

Once employed as an earth scientist, the philosophical demands of the scientific method especially an openness and honesty in coming to a conclusion, can sometimes come into conflict with others. The findings of the geologist may clash with the political and social pressures of government agencies, popular opinion, or the bias and economic rationalism of company superiors. To assist and support earth scientists in maintaining a professional methodology, professional bodies usually have a code of ethics which they offer as guidelines to their members. In addition, employing agencies, such as government private and companies will also have strict codes of professional conduct.

As an example, the following is taken from a former code of ethics of a professional geological society :

Article I - General Principles

1. The practice of geology is a profession, and the privilege of professional practice requires professional morality, responsibility, and knowledge, on the part of the practitioner.
2. Every geologist shall be guided by the highest standards of business ethics, personal honour, and professional conduct.
3. Honesty, integrity, loyalty, fairness, impartiality, candour, fidelity to trust, inviolability of confidence, and honourable conduct are incumbent upon every geologist, not for submissive observance but as a set of dynamic principles to guide a way of life.

Article II - Relation of Geologist to the Public

1. A geologist's responsibility to the public shall be paramount, and he/she shall avoid and discourage sensational, exaggerated and unwarranted statements that might induce participation in unsound enterprises.
2. A geologist shall not knowingly permit the publication of his/her reports, maps, or other documents, for any unsound or illegitimate undertaking.
3. A geologist having or expecting to have beneficial interest in a property on which he/she reports must state in his/her report the fact of the existence of such interest or expected interest.
4. A geologist shall not give a professional opinion, make a report, or give legal testimony, without being as thoroughly informed as might reasonably be expected, considering the purpose for which the opinion, report or testimony is desired; and the degree of completeness of information upon which it is based should be made clear.
5. A geologist may publish dignified business, professional, or announcement cards, but shall not advertise his/her work or accomplishments in a self-laudatory, exaggerated, or unduly conspicuous manner.
6. A geologist shall not issue a false statement or false information even though directed to do so by employer or client.

Article III - Relation of Geologist to Employer or Client

1. A geologist shall protect, to the fullest extent possible, the interest of his/her employer or client so far as is consistent with the public welfare and his/her professional obligations and ethics.
2. A geologist who finds that his/her obligations to his/her employer or client conflict with his/her professional obligations or ethics should have such objectionable conditions corrected or resign.
3. A geologist shall offer to disclose to his/her prospective employer or client the existence of any mineral or other interest that he/she holds, either directly or indirectly, having a pertinent bearing on such employment.
4. A geologist shall not use, directly or indirectly, any employer's or client's confidential information in any way that is competitive, adverse, or detrimental to the interest of employer or client.
5. A geologist retained by one client shall not accept, without that client's written consent, an engagement by another if the interests of the two are in any manner conflicting.
6. A geologist who has made an investigation for any employer or client shall not seek to profit economically from the information gained, unless written permission to do so is granted, or until it is clear that there can no longer be conflict of interest with the original employer or client.
7. A geologist shall not divulge information given him/her in confidence.
8. A geologist shall engage, or advise his/her employer or client to engage, and co-operate with, other experts and specialists whenever the employer's or client's interests would be best served by such service.
9. A geologist shall not accept a concealed fee for referring a client or employer to a specialist or for recommending geological services other than his/her own work.

2.3 Some Famous Geologists in History

Geology, like most modern sciences, has developed by the compilation of knowledge obtained by the keen observations and explanations of men and women who thought clearly and simply about the phenomena they observed.

Some of the most notable individuals who have made contributions to the establishment of the earth sciences have been:

- **Thales** (Greek: 640-546 B.C.) was a mathematician who studied the Nile delta and proposed that water was the basis of all matter.

- **Empedocles** (Greek: 536-470 B.C.) suggested that all things were based on four elements - fire, earth, air and water.

- **Democritus** (Greek: 460-370 B.C.) revised and supported the atomic nature of matter maintaining that all things were composed of small indivisible particles. This idea was unfortunately ignored until the 19th century, when it was revised by **John Dalton** (English: 1766 – 1844), the chemist in 1803.

- **Aristotle** (Greek: 384-322 B.C.) made a classification of all the learning known at that time, including the rudiments of geology.

- **Aristarchus of Samos** (Greek: 310-230 BC) was an astronomer and mathematician who presented the first known model which placed the Sun at the centre of the known universe with the Earth revolving around it.

- **Eratosthenes of Cyrene** (Greek: 276-194 BC) suggested that the Earth was a sphere and calculated its diameter using the angles of the Sun's shadow at two different places.

- **Abd al-Rahman al-Sufi** (Persian: 903-986) Persian astronomer who made the first known observation of a group of stars outside of the Milky Way, the Andromeda Galaxy.

- **Leonardo da Vinci** (Italian: 1452-1519) recognised that landscapes are the results of erosion and that same fossils once lived in the sea. He often incorporated his ideas on river erosion into the backgrounds of his paintings (e.g. the Mona Lisa).

- **Nicolaus Copernicus** (Polish: 1473-1543) proposed the theory that the Earth orbits the Sun.

- **George Bauer** (in Latin: **Agricola,** German: 1494-1556) studied mining and minerals in Germany and wrote a major text book about mining metals (*De Re Metallica*).

- **William Gilbert** (English: 1544-1603) was a doctor who studied magnetism and realised that the Earth had a magnetic field.

- **Galileo Galilei** (Italian: 1564-1642), was an astronomer noted for his improvements to the telescope and consequent astronomical observations. His observations added support for the Heliocentric Theory that the Sun is the centre of the solar system. Galileo has been called the father of modern astronomy.

- **Johannes Kepler** (German: 1571-1630), a brilliant mathematician and scientist who discovered that the Solar System's planets follow elliptical paths, not circular paths and described their motion in a set of laws now known as Kepler's Laws of Planetary Motion.

- **Jean Baptist Tavernier** (French: 1605-1689) was a traveller who collected gems and information about gem-cutting and their trading in the Middle East and Asia, introducing these techniques to Europe.

- **Nicholas Steno** (Danish: 1638-1686) made a detailed study of rock strata and produced the first geological history of an area.

- **Sir Isaac Newton** (English: 1642-1726) was a physicist and mathematician, widely recognised as one of the most influential scientists of all time. Noted for his work on gravity and describing mathematically how the Earth orbits the Sun.

- **Carl Von Linne** (in Latin: **Linnaeus, Swedish: 1707-1778**) was a naturalist who developed the Binomial System of naming living things (including fossils), which is still used today.

- **Nicolas Desmarest** (French: 1725-1815) was a geologist who showed that basalt rock had formed from volcanoes, concluding that this was the mode of formation of other rocks.

- **James Hutton** (English: 1726-1797) was a geologist who proposed the **Principle of Uniformitarianism** i.e. that past geological events were the same as those observed today. He proposed that the earth is active, and that uplifting of the land and intrusion by igneous rocks were caused by internal heat (The Plutonist Theory - also called the **Vulcanist Theory**). Hutton also argued with some elements of the clergy and others, that the Earth was much older than that of 4004 B.C. based upon Biblical calculations given by Bishop Ussher (Irish: 1581-1656) in 1654.

- **William Herschel** (German/British: 1738-1822) was an astronomer who catalogued over 2,500 deep sky objects, discovered Uranus and its two brightest moons, two of Saturn's moons, and the Martian ice caps.

- **Caroline Herschel** (German/British: 1750-1848) was an astronomer and sister of her co-worker William Herschel, with whom she made many significant contributions to astronomy. She became the first woman to discover a comet, and identified several others over the course of her lifetime.

- **Abraham Werner** (German: 1749-1817) was a geologist who proposed that the crystals observed in rocks were formed from evaporation from a primeval sea. His Neptunist Theory, though incorrect, maintained that granites were formed initially, followed by schists, slates, limestones, sandstones, shales and lastly, gravel. During the 18th century the Neptunists were constantly opposed to the Plutonists.

- **Baron George Cuvier** (French: 1769-1832) was a naturalist who proposed a complete classification of animals. He founded the science of palaeontology and explained the sudden disappearance of fossils in certain rocks by certain catastrophes.

- **Friedrich Wilhelm Heinrich Alexander von Humboldt** (Prussian: 1769-1859) was a geographer, naturalist and explorer. He advocated for the long-term systematic geophysical measurement, and laid the foundation for modern geomagnetic and meteorological study.

- **William Smith** (English: 1769-1839) was a practical engineer and surveyor, who studied the rocks and fossils in cuttings whilst building canals and tunnels. He used rocks and fossils for comparing (or correlating) rocks from place to place and mapped the rock layers (strata) of England and Wales. His contemporaries called him Strata Smith.

- **Adam Sedgwick** (English: 1785-1873) studied the rocks of Wales and named rocks of similar form as Cambrian age (after an ancient Welsh tribe). He later named rocks of other ages the Devonian, after the County of Devon, and the Carboniferous age from their coal-bearing feature. These systems or Periods, are now used internationally for rocks of similar age.

- **Roderick Murchison** (Scottish: 1792-1872) continued this system of assigning age names to rocks by coining the named Silurian (after another ancient Welsh tribe) and Permian (after the Perm area of Russia) systems of rocks in Britain. These, too are now used internationally.

- **Sir Charles Lyell** (Scottish: 1797-1875) further developed the science of geology, writing many authoritative text books and giving much support to Hutton's idea of Uniformitarianism.

- **Mary Anning** (English: 1799-1847) was an amateur fossil collector and palaeontologist who became known around the world for her important finds in Jurassic marine fossil beds along the cliffs of the English Channel. Her findings contributed to important changes in scientific thinking about prehistoric life and the history of the Earth in general.

- **Charles Darwin** (English: 1809-1882) was a naturalist who proposed a theory of evolution which led to modern theories and understanding in changes in living things with time. He also advanced a theory of formation of coral reefs.

- **Charles Lapworth** (English: 1842-1920) continued the work of systematically naming rocks by introducing the Ordovician age named after an ancient Welsh tribe.

- **Andrija Mohorovičić** (Croatian: 1857-1936) was a meteorologist and seismologist. He is best known for giving his name to the **Mohorovičić Discontinuity**, the boundary between the Earth's crust and mantle, and is considered as one of the founders of modern seismology.

- **Sir Tannatt Edgeworth David** (Welsh-Australian: 1858-1934) Came to Australia in 1891 from Britain and became well-known for his field explorations of this country and Antarctica. He is remembered for his major work entitled *The Geology of the Commonwealth of Australia.*

- **Sir William Bragg** *(*English-Australian: 1862-1942*)* was a physicist whose work on X-ray devices enabled the structure and dimensions of crystals to be determined.

- **Alfred Wegener** (German: 1880-1930) was a geologist who proposed the hypothesis that the Earth's continents had wandered over the surface of the planet. At first this was opposed, but later evidence so strongly demonstrated his ideas, that the **Theory of Continental Drift** is now one of the major theories of geology.

- **Inge Lehmann** (Danish: 1888–1993) was a seismologist and geophysicist. In 1936, she discovered that the Earth has a solid inner core inside a molten outer core, overturning the idea that our planet's metallic core is entirely molten liquid.

- **Sir Douglas Mawson** (English-Australian: 1882-1958) settled in Australia after coming from Britain and trained under T.E. David. Mawson led the Australian Antarctic expeditions and did much to establish the scientific importance of that continent. Later, he became Professor of Geology of the University of Adelaide.

- **Edwin Hubble** (American: 1899-1953), was an astronomer who determined that the universe is expanding, a concept which later became known as **Hubble's Law.**

- **Harry Hammond Hess** (American: 1906-1969) was an oceanographer and US Naval Officer (Rear Admiral) who is considered one of the founding fathers of the unifying Theory of Plate Tectonics. He is best known for his theories on **Sea Floor Spreading** suggesting that the convection of the Earth's mantle was the driving force behind this process.

- **Jacques Cousteau** (French: 1910-1997), co-inventor of the breathe-on-demand valve for SCUBA diving, Jacques Cousteau led hundreds of marine expeditions and made many documentaries, popularizing undersea studies.

- **Dorothy Hill** (Australian: 1907-1997) was a geologist and palaeontologist, who became the first female professor at an Australian university and was noted for her work on the fossil plants of Australia.

- **S. Warren Carey** (Australian: 1911-2002) was a geologist noted for his early studies into continental drift. He proposed an **Expanding Earth** hypothesis to explain spreading of the sea-floor without the need for subduction zones where oceanic plates are pushed below other plates. This hypothesis has not gained much acceptance.

The work of these great men/women is being carried out today in geological institutions of learning and research. Despite the huge exploration into mineral and energy fields, Geology is still a young science and its techniques are now being applied to the Moon and the other planets of the Solar System.

Chapter 3: Measurement and Accuracy

3.1 Introduction

Geologists, like every scientist, use data obtained from measurement using a variety of scientific instruments. Often large numbers of measurements are taken, using different units of measurements, meaningful and **significant figures** (**SF**), and often very large or very small numbers. In addition, any experimentation which gathers data must also be valid and reliable.

A full explanation of measurement and the statistics used to make sense of the data is beyond the scope of this book, but some simplified ideas of the use of numbers in science are given in the following sections.

3.2 Units of Measurement

These are the name and values of the measurements taken, such as measuring length (name) in metres (unit). To ensure that all scientists use the same units so that measurements can be compared, science uses the common Metric System which was developed after the French Revolution. These metric units are often referred to as **SI Units** from the French term: *Le Système International d'Unités* established in 1960 by the 11th General Conference on Weights and Measures by the *Conférence Générale des Poids et Mesures* (the CGPM) in Paris. This is the official international authority on metric measurement and its units.

Over time, standard units have been established and very precisely defined, so that any country can derive a copy of each standard for their own use in local Standards Laboratories. For example, the most commonly used standard units are the Kilogram (for mass), the Metre (for length) and the Second (for time).

Kilogram (kg) for Mass

Metre (m) for Length

Second (s) for Time

Various prefixes are used to help express the size of quantities which could be extremely small or large (e.g. a millimetre = 1/1000 of a metre and a kilometre = 1000 metres).

PREFIX	SYMBOL	POWERS of TEN	ORDERS of MAGNITUDE
yocto	y	0.000000000000000000000001	10^{-24}
zepto	z	0.000000000000000000001	10^{-21}
atto	a	0.000000000000000001	10^{-18}
femto	f	0.000000000000001	10^{-15}
pico	p	0.000000000001	10^{-12}
nano	n	0.000000001	10^{-9}
micro	μ	0.000001	10^{-6}
milli	m	0.001	10^{-3}
centi	c	0.01	10^{-2}
deci	d	0.1	10^{-1}
	unity	1.0	10^{0}
deca	da	10	10^{1}
hecto	h	100	10^{2}
kilo	k	1000 (Thousand)	10^{3}
mega	M	1000000 (Million)	10^{6}
giga	G	1000000000 (Billion)	10^{9}
tera	T	1000000000000 (Trillion)	10^{12}
peta	P	1000000000000000	10^{15}
exa	E	1000000000000000000	10^{18}
zetta	Z	1000000000000000000000	10^{21}
yotta	Y	1000000000000000000000000	10^{24}

Table 3.1: Some of the most commonly-used prefixes

3.3 Scientific Notation

When very large and very small numbers are used, scientists use the **scientific notation** to refer to these numbers by expressing them as powers of ten. For example:

$$0.001 = 10^{-3}$$
$$0.01 = 10^{-2}$$
$$0.1 = 10^{-1}$$
$$1 = 10^{0}$$
$$10 = 10^{1}$$
$$100 = 10^{2}$$
$$1000 = 10^{3}$$

This can be used for any number with it being expressed as a power of ten and a multiple e.g.

65331290 would be written as   **6.5331290 x 10^7**

and **0.0000256** would be written as   **2.56 x 10^{-5}**

Notice that these numbers are always expressed as a digit (e.g. 2) which is then followed by a decimal point and then the rest of the number (e.g. .56). This gives the <u>stem</u> which is then followed by the <u>power of ten</u> (e.g.10^{-5}). For large numbers, this power is a positive value and is the number of digits after the decimal point (i.e. 7 in 6.<u>5331290</u>). For small numbers, the power is a negative value (being tenths etc.) and is the number of digits behind the decimal point <u>including the first digit used</u> (i.e. in 0.<u>0000</u>256 there are four zeros + the <u>2</u>).

When referring to the order of magnitude, i.e. the approximate powers of 10 that a number has, only the number of the power e.g. 10 to the 7 or ten to the minus 5 is given. Also, in general estimation only the prefix may be used e.g. mega.

In calculations, the stems (e.g. the 2.56 in the example above) are used as in normal decimal operations and the powers of ten are applied separately such that in:

Multiplication   -   <u>add</u> **the powers; and in**
Division   -   <u>subtract</u> **the powers**

For example: Consider the calculation   **6947   x 124**

This becomes **(6.947 x 10^3)**   x   **(1.24 x 10^2)**

or **(6.947 x 1.24)** to the power of **(10^3 + 10^2)**

or **8.61428   x 10^5**

More realistically, this may be rounded to **8.61 x 10 ^{5}**

3.4 Accuracy

As in all sciences, geologists strive for the best accuracy possible in their measurements. This not only depends upon their powers of observation, but also the nature and construction of the instruments which are used. Accuracy concerns how close a measurement is to the real value. This is always in doubt as the instrument used will have an error due to its limitations of construction.

Instruments can be analogue or digital, with the latter becoming more common than the former. With analogue instruments, all of the scale of measurement can be seen and the user must judge the position of the measurement on the scale. With digital instruments, the reading of the measurement only is shown on a screen as a series of digits. As a general rule the degree of accuracy of the instrument would be:

- For **analogue instruments**, the error of accuracy would be <u>half</u> of the smallest unit on the scale because it is assumed that an observant person can accurately estimate the halfway point between two marks.

For example, consider the scale shown below. It is measured in hundredths of a gram (i.e. there are ten units between 0.9 g and 1.0 g):

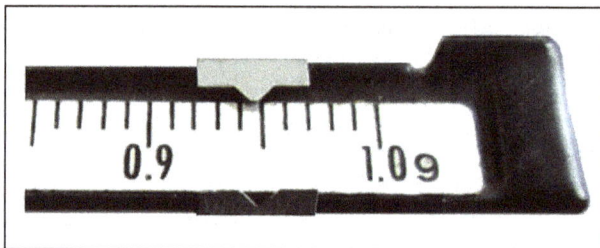

Figure 3.1: A scale of a gold balance

The instrument error here is half of the smallest unit (i.e. 0.01 gram) which is:

0.005 grams

This means that the measurement value from this scale would be:

0.940 +/- 0.005 grams

Notice that whilst the measurement must include the 0.94 units, it must also include an added zero to show the significance of the error. Also that as this error value is an estimation, it is taken as a plus or minus value. What is being said here is that the measured value is between 0.940 and 0.945 grams.

- For **digital instruments**, this error would be the smallest unit shown, because this all that can be observed on the screen.Consider a digital instrument such as the following electrical meter measuring the electrical resistivity of a specimen of galena (lead sulfide):

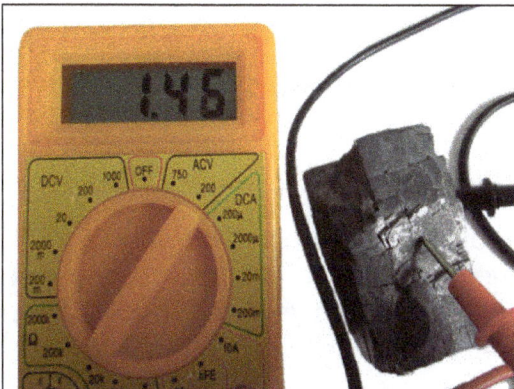

Figure 3.2: A multimeter measuring the resistivity of the mineral galena

Here, the <u>only</u> value that can be seen is on the screen – is 1.46 ohms. The error here would be +/- 0.01 of an ohm, so our measured value would be:

1.46 +/- 0.01 ohm

Usually a scientist will not just take one measurement, but as many as is practically possible. This is because there are other factors which are involved in every measurement which could cause errors, such as **environmental errors** e.g. the lighting, wind and temperature as well as the human observing ability of the scientist, or the **operator error**. These factors are often called **random errors**. There may also be an **instrument error** due to the poor calibration or the setting up the instrument scale and the condition of the instrument itself such as its battery charge and cleanliness. Any other variable which can be avoided or removed is called a **systematic error**.

In the case of the digital measurement of resistivity (above), the scientists would   take several readings, which might be:

1.46; 1.44; 1.50; 1.45 and 1.43 ohms

Each having an instrument error of +/- **0.01 ohm**

A simple average of these values would give **1.458** ohms, but this value seems more accurate because it is now to three decimal places instead of the two derived from the instrument, so it is incorrect to state this value because it is unrealistic to do so. Instead, one should use the same number of figures as given by the instrument as a real estimation of the truth of the measurement. In this case, as the measurement is almost to the next hundredth, one can round the value up if the last digit is more than 5. If the last digit had been less than five (e.g. 1.453), then the number would be rounded down. If this last value is equal to 5, then it would be best to consider the worst error possible and so as a precaution the value would also be so rounded down.

In this example, the final average would be rounded up to a significant figure of 1.46. However, before finally stating the measurement, one must remember that there is also an instrument error of +/- 0.01 for each of the five measurements. In reality the total error for this final average is added i.e.5 x 0.01 or 0.05, so the final measurement would be given as:

1.46 +/- 0.05 ohms.

If different instruments each with a different accuracy were used, then one would follow the same procedure but must quote the final value to a significant figure which will represent the <u>least accurate</u> of all the readings. The final error would still consist of the addition of all of the errors.

$$1.46 \quad +/0.01;$$
$$1.462 \quad +/-0.001;$$
$$1.500 \quad +/- 0.005$$
$$1.420 \quad +/- 0.001$$
$$1.4400 +/- 0.0001$$

Here, the simple crude average of the values is **1.4564** and the addition of the errors   is **0.0171**. None of these figures mean very much in scientific terms, so one must quote the value to its smallest number of significant figures i.e. three significant figures due to the limitation of the 1.46 measurement, and use a rounded value of the added errors to match this significance:

1.46 +/- 0.02 ohms.

For greater accuracy, scientists may use the **Root Mean Square** (**RMS**) of these values. This is the square root of the sum of the square of the values. For example, consider these values (above) and their errors:

The RMS value would be:

$$\sqrt{(0.01)^2 + (0.001)^2 + (0.005)^2 + (0.001)^2 + (0.0001)^2}$$

or **0.03 ohms**

In calculations, when data values are added or subtracted, the errors involved in each datum must also be added. Those errors due to addition and subtraction in which the same units are used, are generally referred to as **absolute errors.** However, if the units of measurement are different, or if multiplication or division is involved, then the error will be expressed as a **relative error** by using a relative percentage rather than some combination of different units. For example, in calculating the density of a mineral where:

38

Density = Mass / Volume

the final value will have an error expressed as the addition of the percentage errors for each measurement. That is, the absolute error of each measurement will be calculated as a percentage of the main value, and then added together as common percentages to give the final error of the calculation. For example, consider the following calculation:

Mass of Mineral = **234.05 +/- 0.01** grams
Volume of Mineral = **52.1 +/- 0.1** millilitres (= 52.1 cubic centimetres)

Now Density = Mass / Volume or

= **234.05** grams
  52.1 cubic centimetres

= **4.49232245681** etc. etc. (according to the calculator!)

As different units are used, the percentage errors are found for each value's error:

0.01 as an error of **234.05** is **0.01/234.05 x 100 = 0.4%**

and **0.1** as an error of **52.1** is **0.1/52.1 x 100** = **0.2 %**

The here, the relative error would be the addition of the calculated percentage errors or:

0.4% + 0.2% = 0.6%

Therefore, the correct value for this density (using the appropriate i.e. lowest significant figure) would be:

4.49 +/- 0.6%   grams/cubic centimetres (g/cc)

Note that as the minimum number of significant figures in the problem i.e. of the 52.1 value, are only three, so only this number of figures will be used in the answer, <u>not</u> the number of decimal places. Also, the unit of measurement must not be forgotten, and this should be given as the appropriate combination of units or a specific unit if one has been defined i.e. g/cc for grams per cubic centimetre in the example above.

In summary, to ensure that the values which are given are significant, the following basic rules should be remembered so that the numbers given for any scientific measurement and calculation are realistic and reflect the actual limitations of the instruments and their accuracy. The basic rules are:

- **All non-zero digits are significant** e.g. so 235 has three Significant figures and 9.886 has four S.F.

- **Zeros between non-zero digits are significant** e.g. 406 has three S.F. and 4.009 has four.

- **Leading zeros are never significant** e.g. 0.00024 has only TWO S.F. and 0.0204 has three S.F.

- For a number with a decimal point, **trailing zeros (i.e. those to the right of the first non-zero digit), are significant** e.g. 2.7800 has five S.F. because the trailing zeros reflect the degree of accuracy of the instrument which can give values to this many decimal places and

- For a number without a decimal point, **trailing zeros may or may not be significant** and there will be an uncertainty about the validity of giving such a value. More information on the instrument used and the calculations are needed to clarify the significance of trailing zeros e.g. 34.8000 may mean a very accurate reading of six S.F. or that the individual has not rounded off a calculation for an instrument which only reads to a lesser number of significant figures (say 34.8).

3.5 Validity and Reliability

All scientific studies, and especially measurement under laboratory or field conditions must be appropriate to what is being studied and it must be able to be reproduced at other times and by other Scientists.

The **validity** of any observation, experiment or gathering of data, is how fairly or honestly the results tests the hypothesis in question. In a valid experiment, all the variables i.e. the influences that would affect the experiment, are kept constant or are removed, apart from those being investigated. All systematic errors have been eliminated and random errors are reduced by taking the average of multiple measurements. For example, using mathematical stream-flow equations to calculate sediment flow in northern rivers of Australia would give consistent and reliable results every time at that locality, but the conclusions from this investigation would be wrong if the mathematical modelling came from (say) European sources where the stream environments would be different. In other words, the investigation of the scientist should match the criteria need to test the hypothesis proposed.

The **reliability** of an experiment or measurement is its ability to be repeated and show some consistency in its data and calculations. If an experiment is repeated many times, it will give similar results if it is reliable. In referring to the data of others, reliability refers to the account of the source as being trustworthy. Of course, reliability depends upon who provides the data and how it is communicated. For obtaining accurate data from others, earth scientists prefer to go to the original source which may be other colleagues, or to refereed journals of good repute. For example, one would believe data from a well-known scientist who has written in a national journal of a scientific body (e.g. the Proceedings of the Royal Society, London), rather than some unknown's ramblings on an internet blog, although these may be useful as indicators rather than scientific evidence. The reliability of any investigation, journal or internet site can be assessed by comparing it to several other similar sources on the same topic.

Chapter 4: Into the Field

4.1 Introduction

In some respects it is a retrograde viewpoint in a world dominated by electronic devices, satellite geology and computer modelling, to describe the simple techniques of going into the field, navigating across land and making maps from the data taken. A modern field geologist can obtain a large amount of data about a proposed exploration site without leaving the office, and much of this data can then be used to computer model the site, or at least produce maps. However, it still requires a human to explore the area and verify the electronic data before positive decisions are made about the actual geology of the area and is usefulness.

Since the launch of the Earth resources satellite LANDSAT 1 on July 23rd, 1972, there have been over 130 such satellites launched from several different countries – one of the last being BISONSAT, a training satellite for measuring ground cover and cloud types, on October 8th, 2015. There have also been almost 200 communications satellites enabling radio, television and navigation services worldwide. Earth resource satellites have been used since the 1970's to provide imagery ranging from simple satellite photographs to more sophisticated and high-resolution thematic mapping. Satellites can measure and transmit data with some degree of accuracy for such uses as gravity, geomagnetic and radiometric surveys as well as being able to penetrate vegetation and snow cover with infra-red light and the depths of oceans with certain **LASER** wavelengths.

Figure 4.1: Early LANDSAT1 photo of the Californian coast 1972 (Photo: NASA-JPL)

Figure 4.2: False colour image of the Mississippi Delta compiled from several images taken (Photo: NASA/Earth Observatory)

In the field, the geologist may use a hand-held **Global Positioning System (GPS)** to navigate across the land, taking wave-points to plot position, and then recording the findings and measured data electronically using a **Geographical Information System (GIS).**

Figure 4.3: Artist's impression of the latest Polar-orbiting Satellite for weather and oceanographic research (Photo: NOAA).

Additional storage and processing of this data can also be entered into a field **Personal Data Assistant (PDA),** a hand-held personal computer designed for rugged conditions. Later this data can be uploaded directly to the office via satellite. It is still a good idea to have some of the old skills of field navigation and map-making, as well as understanding how to move across country with accuracy of observation and how measured data is put together to make geological maps.

4.2 Basic Approach

One usually has good reason to go into the field. It may be only a casual exploration searching for things of interest such as gems, limestone caves, fossils or mineral specimens, or it may be a full-scale geological exploration.

For a thorough geological exploration, professionals usually follow a sequence of research methodology involving:

1. Clarification of objectives - what is the research about? What is to be achieved? These may be the choice of the researcher or they may be a project of a private company, research organisation or government institution but whoever initiates the research would expect a clear statement of objectives, especially if funding is to be raised.

2. Review of previous research - before the preliminary plans are made, past research, leases, publications and the work of others in the planned site and of the targeted resource should be reviewed. This might be a simple Internet and literature search, discussion with any former researchers or a full investigation including formal review of all past research of interest and a search of government documents, land-ownership and access and former and existing mining and other leases. This would also be a time of gathering as much data as possible about the site and of any target resource. This might mean the use of any aerial or satellite imagery or other data such as magnetic, radiometric and gravitational surveys.

3. Preparation and planning - can be started only when a clear idea of objectives and all aspects of the target area have been clarified. These aspects will govern such factors as:

- Access to the area including the crossing of private land, traditional rights of local people, degree of difficulty or terrain, transportation to and across the area, and possible base camps and support.

Figure 4.4: Whenever possible, the sacred sites and rights of indigenous people should be respected.

- Length of time estimated for the exploration, including travelling to the site, setting up camp, exploration methods and of most importance climatic and other limitations.

- Equipment needed for sampling and also for basic living conditions. This will be determined by the nature of the site area, the possible climatic conditions and any sampling, testing and data acquisition which can be done in the field.

- Communications and resupply may be a major consideration if the site is remote. The use of off-road vehicles and even helicopters may be a consideration and

Figure 4.5: Some research sites are in difficult locations - Estacion Cientifica Almirante Brown, an Argentinian research base at 65° South, Antarctica.

- Personnel needed for general transportation, camp establishment, field exploration and testing should be considered. Generally field staff would be kept to a minimum and usually it would only involve the geologist and any support staff needed.

4. Execution of the research involving the field exploration, sampling and data gathering within the field by the geologist or research team. This might be supplemented by more sophisticated field exploration such as specific drilling and drill core logging using geophysical techniques.

5. Post-field analysis of the data collected with further research into new discoveries. This may also involve the correlation with other related research, computer modelling and construction of surface and sub-surface geological maps and

6. Final conclusions and recommendations derived from the study. Usually this will be in the form of a written report to the parent organisation and if the findings are of major scientific or public importance it may also be published in an appropriate refereed scientific journal.

4.3 Going into the Field

Going into the field can be a dangerous activity, especially when going alone, which is what many geologists often do. Having previewed all the maps, aerial photographs and satellite photographs available for the area, the geologist would then most likely decide on the type of **traverse** or route to be taken across the area which would give the best chance of obtaining the maximum amount of valid data. If the exploration also involves field mapping, then an appropriated base map should have been drawn prior to arriving at the site. This base map will normally be done back in the laboratory from existing images and maps, and it will:

- Cover the area of the research.

- Be drawn to scale using any available **topographical maps** (the common land surface map) updated from aerial and satellite images.

Figure 4.6: Pocket stereoscope and paired satellite photos which overlap and give a 3D image.

- Show the major topographic features such as relief i.e. height above sea level, rivers, geographical coordinates such as detailed Latitude and Longitudes – usually in degrees, minutes and seconds, grid and magnetic north, the magnetic variation for that specific location, and any information regarding access such as roads, navigable rives and potential helicopter landing zones and

- Show possible lines of traverse or routes which may be used in exploration.

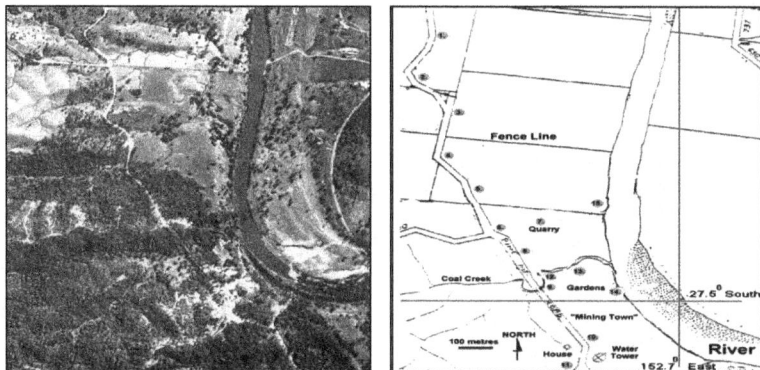

Figures 4.7 & 4.8: An aerial photo (left) and the small base map drawn from it (right) showing numbered data sampling points along a road.

This map could then be used for any appropriate survey method used in the field. This could vary depending upon the scale and accuracy of the intended project. It may be a simple exploratory investigation by a single geologist, or it may involve a full-scale survey using a team who would map the area using the best survey equipment available. This might include LASER **geodometers** with GPS base locations and other survey instruments for accurate plotting of all geological formations and beds. Detailed photography, both on the ground and by remote aerial vehicles (drones) or helicopter, could also be used to support the surveying methods.

In the field, the geologist will walk along a planned route or traverse which will provide the best search pattern of the target area and for the collection of data and specimens. Traverses could be either open or closed, depending upon the complexity of the task and the amount of ground that the Geologist wishes to cover.

Open Traverses have a separate start and finish and can be of several types:

- Simple Linear Traverse is one by which the field geologist walks along a designated track, taking measurements at each new change in the geology or land surface. The route for this traverse may be along a road, rail cutting, gorge, ridge, stream, across country following a specific geological formation, or a compass bearing which crosses geological boundaries. At various points along the traverse, the geologist might make short trips.

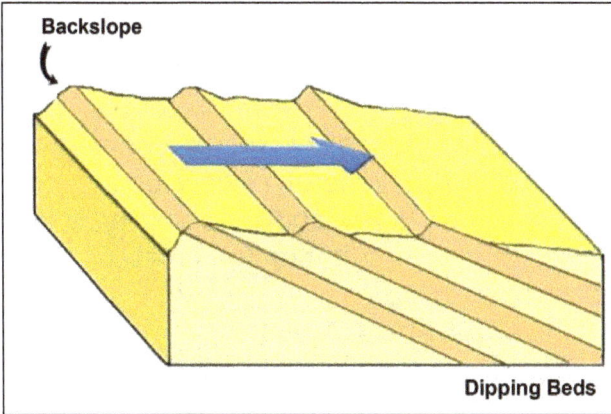

Figure 4.9: A simple linear traverse across dipping beds

- A Stream Traverse is a type of linear traverse up or
down a stream with water sampling at each tributary
or spring. Looking for specific minerals, the geologist
would take water samples at every new tributary
whilst walking along the main stream. Whilst going
downstream is usually easier, walking upstream
ensures that every stream is located and sampled.

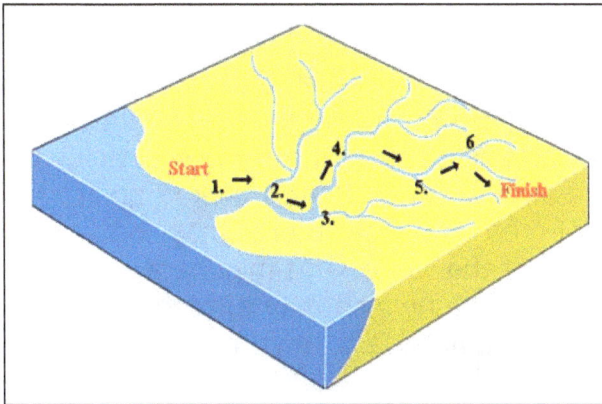

Figure 4.10: Diagram showing a stream traverse for water
sampling

- Radiating Traverse - is a simple search around a particular point of interest. Here, the geologist systematically goes out from a central base along designated directions as a set of compass bearings. This is ideal for a general review of an area rather than a detailed survey and mapping project. This is the type of search suitable for amateur prospectors around a base camp.

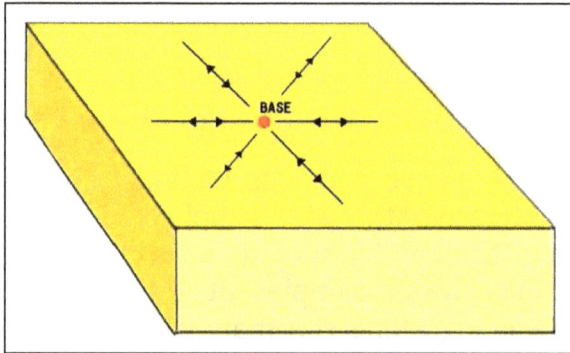

Figure 4.11: Diagram showing a radiating traverse for simple exploration from a base camp

- **Closed Traverses** are useful for covering a wider, specific area of interest and has the advantage of giving a more complete picture of that area. It also starts and finishes at the same location, usually the base camp or a designated pickup point. Along each leg of the traverse there may be something of interest of the route. This may be an outcrop of rock, a stream gulley which may expose rock, a change in the soil on a hillside or anything worth investigation:

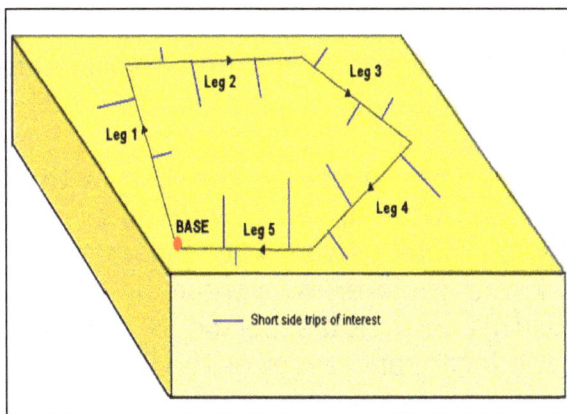

Figure 4.12: A closed traverse showing several legs and tangential side trips

4.4 Safety in the Field

Safety is one aspect of earth science field work which cannot be emphasised enough for both professional scientists and the amateur. Every year the media will have stories of people who have ventured out into rugged countryside and have fallen foul of the terrain or the weather and suffered because of their lack of preparation and personal fitness. In some parts of the world, the weather conditions and terrain are dangerous at all times. Other places are deceptive, in that they seem to be suitable for a casual exploration but then suddenly become dangerous because of a sudden change in the weather or a sudden geological event such as an earthquake, volcanic eruption, landslide or cave collapse. When going into the field, one must take certain precautions, the scale of which will depend on the length of the trip, the number of people, the terrain, weather conditions and the nature of the task. Considering only the personal safety of an individual, the following suggestions are worth noting:

- Have a definite plan for the trip with direction, distance and time of return written down and given to a responsible person who can arrange a search party if the explorer is overdue by a certain time. Usually this responsible person is someone back at base, the local park ranger or the local police.

- Personal clothing and equipment must be suitable for the climate and terrain. Good boots, clothing which covers the body, arms and legs, and a hat are essential. In hot weather the clothing can be lightweight but there should be no compromise about the footwear. Always assume that it will rain or become colder. A wet-weather covering or compact padded jacket should always be taken into the field, in a backpack if necessary. An ex-army hootchie (or tents – half shelter) with a hood, is good for bad weather as a cape or to use as a shelter.

- Water and food are essential, even for a day trip. Accidents happen and one should carry water and food for at least two days. In hot, dry conditions more water will be needed, and in places where water is scarce, an adequate supply should be carried in water bags. Where streams are likely to be encountered, they may not be suitable for drinking because of bacterial, worm or chemical contamination. Filters and chemical sterilizers are often a good backup to boiling water for several minutes for coffee or tea. There should also be enough water to reconstitute any dehydrated food that is carried. This food is lightweight and a great alternative to heavy tinned food.

- Shelter is a consideration in places of extreme climate or where the weather is likely to suddenly change, mountainous areas for example. The army half-shelter mentioned above makes a good, temporary shelter from cold wind, snow and hot sun. For overnight trips, a good quality alpine-style lightweight sleeping bag is essential. An extra strip of orange plastic sheet about two metres by about 1 metre is useful as a ground sheet – two of these are good as distress panels for air rescue. The **wind chill factor** in exposed areas is important. A slight wind will drop the temperature down quickly, especially if the weather is already cold and wet.

```
SPEED                          TEMPERATURE
(km./hr.)                         (0C)
-------------------------------------------------------------------
Calm  5   2  -1  -4  -7  -9 -12 -15 -18 -20 -23 -26 -29
  8   1  -1  -4  -7  -9 -12 -15 -18 -20 -23 -26 -29 -32
 16  -1  -7  -9 -12 -15 -18 -23 -26 -29 -32 -37 -40 -43
 24  -4  -9 -12 -18 -20 -23 -29 -32 -34 -40 -43 -45 -51
 32  -6 -12 -15 -18 -23 -26 -32 -34 -37 -43 -45 -51 -54
 40  -9 -12 -18 -20 -26 -29 -34 -37 -43 -45 -51 -54 -59
 48 -12 -15 -18 -23 -29 -32 -34 -40 -45 -48 -54 -57 -62
 56 -12 -15 -20 -23 -29 -34 -37 -40 -45 -51 -54 -59 -62
 64 -12 -18 -20 -26 -29 -34 -37 -43 -48 -51 -57 -59 -65

No Extra     LITTLE        INCREASING DANGER - FLESH       GREAT
Effect       DANGER        MAY FREEZE IN 1 MINUTE          DANGER -
above 64                                                   FREEZING
                                                           IN  30 SEC.
```

Table 4.1: Chart showing the wind chill factor data for lower temperatures. For example, in wind of 8 km/hr, a still air temperature of 5°C will feel like 1°C and at 16 km/hr. it will feel like -1°C.

- Navigation equipment is essential even for short tips and especially in very close vegetation or mountainous country. Most geologists will have a good compass and probably a GPS unit but these, like all electronic devices, have their limitations and are subject to failure at the wrong time. A good understanding of the general area, some local advice and an escape plan should an accident happen or the weather changes, is also a wise idea.

4.5 Observation in the Field

A good geologist is an observant one. A good knowledge of geology is one thing but observing slight changes within the environment and then relating them to geological knowledge is what is needed. In most cases, nature is essentially random, so with practice, the subtle changes which indicate a change in the rock type and orientation of rock beds or changes to the land surface will become apparent. For example some features which may indicate a change in the geology include:

- Obvious outcrops of rock which may be a different shape or colour to the general surroundings. Volcanic structures, such as the remains of an ancient volcanic pipe or **neck** or vertical **dykes** and horizontal **sills**, are often very prominent because of their sudden interruption of the landscape. Sharp limestone pinnacles and sandstone cliffs are also abrupt and obvious.

- Rounded boulders or **tors** on relatively flat plains indicate that there is a large body of igneous rock, called a **batholith**, of granite or similar rock below. Sometimes these intrusions harden the surrounding rocks by **contact metamorphism** and then the intrusive rock weathers more quickly to give a large, rounded valley.

Figure 4.13: Tors on the surface indicate a large igneous structure such as a batholith below.

59

- Changes in soil colour often indicate the nature of the parent rock below. This is especially true when there is a weathered ore body reaching the surface and the top layer forms a hard, raised **gossan.** This is usually of a reddish colour due to the iron oxides looks like a small rounded cap.

- Changes in vegetation are often associated with changes in soil type derived from the rock below. Some vegetation becomes stunted when they lack certain mineral nutrients, and these can often show the boundary between, say a sandstone with a lime cement (good) and a poor soil having only sand. Observing from a high vantage point or using a local aerial photograph may show such boundaries as distinct curved or linear features. The author once located small, isolated patches of an ancient lava flow which had dripped over the edge of an escarpment by looking for banana plantations which the locals had grown on the resulting rich basaltic soil.

- Sudden changes in terrain such as sudden changes in the direction of rivers, a long escarpment or sunken valley, often indicates fault lines caused by earth movements. Any linear feature, especially when seen in large scale from aerial or satellite images, denotes something of interest.

Figure 4.14: The sudden rise of the mountains along New Zealand's wide Canterbury Plain marks the start of the Transalpine Fault

Chapter 5: Rocks and their Features

5.1 Introduction

Rock types, the soils which come from them, and the geological structures formed by rocks or land movement, are amongst the features which geologists look for in the field. It is good to know of some of the basics about rocks and their features. In time, a geologist will get a feeling for a particular landscape and will be able to look for specific indicators. Usually, however, it is the unexpected shape, colour or texture of a surface which will attract attention. This applies not only to rocks, but also to the soil, vegetation and sometimes climatic effects.

5.2 Rocks as Indicators

Some basic features of the three main types of rocks are useful in field identification. These are:

- **Igneous rocks** are usually crystalline as they have been formed when molten rock material has cooled. Large crystals are formed if the rock has formed deep underground as an intrusive igneous rock from molten material called magma. Igneous Rocks that have small crystals, or none at all, have been suddenly cooled on the surface as extrusive igneous rock, or volcanic rock, from molten lava. Igneous rocks are relatively hard, but they also weather very quickly. Ancient volcanic landforms are often obvious by their sharp peaks of old volcanic **necks** or from flat, lava-capped hills of old lava flows. These flows may also have many six-sided shapes due to columnar jointing which formed as the lava cooled.

Figure 5.1: The flat, escarpment of the basalt plateau at Antrim, Northern Ireland.

Figure 5.2: The volcanic structure of Ball's Pyramid with Lord Howe Island beyond, eastern Pacific.

- **Sedimentary rocks** are formed in horizontal layers called strata, deposited from fine to coarse sediments by wind, streams, oceans or glaciers. Volcanic ashes can also fall and form strata of fine material. Strata which may have been originally horizontal, may later be tilted, folded or faulted by earth movements.

Figure 5.3: Sedimentary strata tilted in a coal pit near Ipswich, Queensland, Australia,

Figure 5.4: Horizontal strata of volcanic ash at Mt. Gambier, South Australia.

Limestone, a sedimentary rock formed from buried ancient coral reefs, weathers chemically and often produces sharp pinnacles and areas of deep caves, gorges, cliffs and rectangular jointing. The surface of such landscapes is called **karst topography.** The rock is usually a dark grey colour with lighter grey and brown weathered patterns. Karst country is very apparent by its rough, often dry and barren surfaces. Caves can be located in colder weather by the small clouds of warmer, moist air which they emit and sometimes by large trees which often put their roots down into the moist crevices.

Figure 5.5: The Barren, southwestern Ireland, a bare deforested coastal plain of limestone karst.

Figure 5.6: Caves in eroded limestone at Jenolan Caves, New South Wales, Australia.

- **Metamorphic rocks** have been formed from previously-existing rocks which have then been changed by heat from nearby igneous bodies or great pressure due to mountain building processes. These rocks can form large mountain ranges, but are less easily discovered if they are old and eroded to a flat plain or rounded hills. **Contact metamorphic** rocks such as marble, quartzite and hornfels, are formed from heating by a nearby igneous body, and are usually harder and smoother than their parent rocks. The zone of heating or **aureole**, which contains the metamorphic rock may only be only a few centimetres if it was formed by a small igneous intrusion such as a dyke or sill, but it could be several kilometres wide near a large batholith.

Figure 5.7: The Samford Valley northwest of Brisbane, Australia, contains an old, weathered batholith which has eroded relative to the harder surrounding metamorphic rocks.

Regional metamorphic rocks such as schists and gneisses have been formed over a very large area by extreme pressures within relatively new mountain ranges. These may be greatly eroded over long periods of time to rounded hills, but the rocks and soils here, may still contain shiny, flattened crystals of the mica. Stream sediments will often contain flakes of mica which reflect well in sunlight and look like specks of gold.

Figure 5.8: Shiny grey schist eroded to the front of the Franz
Josef Glacier, South Island, New Zealand.

Chapter 6: Landforms

6.1 Walking in the Field

In some cases, the geologist may need to go off the road and walk across country to obtain information about the local geology. A good knowledge of how to read the terrain and how to cross it is vital. The surface of the land or its topography, will determine the routes that the geologist will take in conduction any exploration as well as any engineering considerations which will have to be made in any major project which is to be conducted.

The difficulty of walking across country depends upon the:

- **Gradient** or the steepness of the slope, given as a ratio of the vertical height achieved on the slope to the distance covered

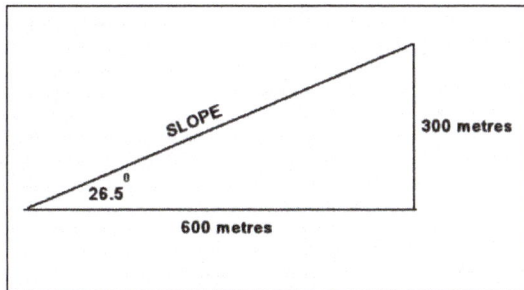

Figure 6.1 Gradient of a slope

horizontally on the map e.g. a gradient of 1:2 means that the walker has covered 600 metres of horizontal distance whilst climbing up 300 metres vertically.

The gradient can be estimated quickly by looking at the angle of the slope or using a Brunton Compass side-on to measure the inclination.

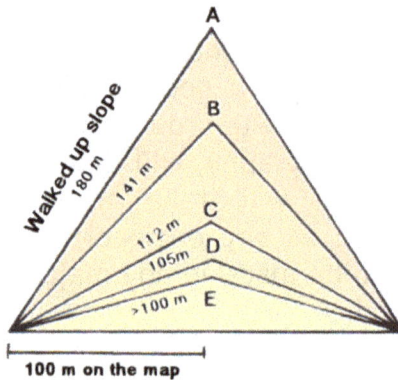

Figure 6.2: Gradient triangle showing actual distances walked for 100 metres on the map (Letters refer to the table below)

Some angles for common gradients are:

1:1 = 45 ^{0} a very steep climb
1:2 = 26.5^0
1:3 = 18^0 a steep hill
1:4 = 14^0
1:5 = 11^0
1:20 = 3^0 easy cycling

KEY	GRADIENT	DISTANCE		PACES WALKED			
		Map	Walked	Uphill		Downhill	
				Number /100m	Length cm	Number /100m	Length cm
A	1.5:1	100	180	540	30	360	45
B	1:1	100	141	282	45	212	60
C	1:2	100	112	168	60	149	68
D	1:3	100	105	140	68	126	75
E	1:5	100	>100	120	75	120	75

Table 6.1: Showing the actual paces walked and the length of stride for various gradients (length refers to the length of a typical pace in centimetres)

As a "Rule of Thumb", for relatively open country, one should allow approximately:

ONE HOUR for every 4 km. measured on the map
plus
ONE HOUR for every 450 metres of vertical height climbed.

Time of travel will also depend upon:

- Vegetation cover will impede the walker if it is close cover, especially if it contains vegetation with thorns or vines. As a rough guide the following estimations of distance on a flat surface per hour are suggested:

Open Country/Track	4km.
Open Forest	1-2km.
Dense Rainforest	0.5-1km.
Tall Grass	0.5-1km.
Bracken Fern	0.2-0.5km.
Swamp (will vary)	0.1-0.2km.

- Condition of the ground surface which may be slippery with mud and snow which will reduce the pace to about swamp pace on the list above.

- Weather conditions, with rate being reduced in rain, snow and extreme heat. In extreme conditions it is best to shelter and wait. Visibility is also greatly affected by weather conditions and

- Fitness of the slowest member of the party will limit any team exploration.

6.2 Topographical Maps

Topographical (or survey) maps are a representation of the land surface and most of the features found on it. Heights above sea level (a.s.l) are usually shown by **contour lines**, which are lines joining places of equal heights, but some simpler maps may only have colour shading to represent approximate relief. Whilst satellite **thematic mappers**, GPS handsets and similar electronic devices tend to make the navigation by topographical map redundant, it is still useful to understand some of the basics of using maps. They are most useful when a base map needs to be drawn prior to a major field exploration.

1:50 000 Topographical Map   UTOPIA

SCALE 1:50 000

METRES 1000   0   1   2   3   4   5 KILOMETRES

CONTOUR INTERVAL 10 METRES

The representation of a road on this map is no evidence of the existence of a right of way

Figure 6.3 An example of a topographic map showing the main features

The main features of these maps worth understanding are the:

- **scale** which is a representation of the horizontal distance on the map. This is often shown at the top of the map as a ratio of map value to actual length on the land. The Utopia map (above), is a 1 in 50,000 scale map, meaning that one centimetre on this map represents 50,000 centimetres, i.e. 500 metres, on the ground. This scale is also given in a more useable form at the bottom of the map as a linear scale;

- **map Legend** is usually at the bottom of the map and shows the symbols for the features on it such as vegetation, places of inundation, roads, buildings and the like. Roads and rail lines with cuttings through rock, and quarries are of special interest to the geologist;

- **grid squares** are the squares on the map made by the intersections of the vertical lines called **eastings** and horizontal lines called **northings**. These are used for distance estimation and for locating positions on the map. They are also useful when communicating about specific points on the map to another person who has the <u>same map</u>. If giving a location from the map to another person, especially a rescue team, who probably would not have the same map, then the geographical bearings should be given as international latitude then longitude. Longitude (vertical) and latitude (horizontal) are also shown along the edges of the map. When giving a **grid**

reference, use the numbers along the bottom of the map first (i.e. the eastings) and estimate subdivisions to a tenth place. There are reference cards or **romers** for each map scale which can be used as a ruler to help in this estimation. Then do the same for the vertical numbers (northings). For example, on the topographical map shown in figure 6.3, the location of the end of the jetty at Port Utopia would be given as **796584.**

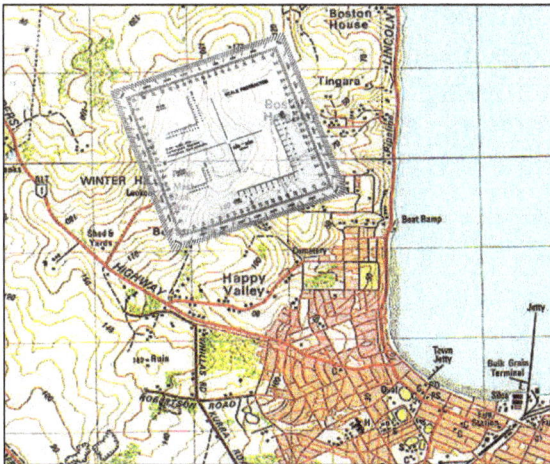

Figure 6.4 A Map protractor with romers

- **Contour lines** are lines joining places of equal height and often are shown as brown or black lines. The vertical distance between them is called the **contour interval**, and this can be found by comparing the numbered heights of two adjacent contour lines. In the following map, the contour interval is given as 10 metres.

- **North points** are shown at the bottom right of the map and indicate the direction of various representations of north. These are **Grid North (G.N.)** which simply shows direction of the vertical, numbered grid lines on the map; **True North (T.N.)**, which points towards the Geographical North Pole; and the **Magnetic North (M.N.)**, which points towards the North Magnetic Pole and is the direction that a compass needle will point if placed on the map. The angle between True North and Magnetic North is called the **magnetic declination (or variation)**. The angle between Grid North and Magnetic North is the **Grid Magnetic Angle (G.M.A.)**. Both declination and G.M.A. are specific for that map and locality.

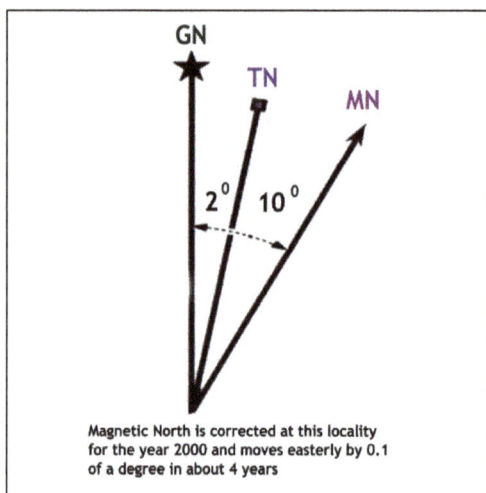

Figure 6.5: North points which may be shown on a map

Simple calculations must be made so that a magnetic compass can be used to follow a bearing plotted on a map which uses grid north. Bearings (or directions) measured on the map with a protractor must then be converted to a magnetic bearing so that the compass can be used to follow this direction on the land. This is done by calculating the grid magnetic angle by referring to the diagram of north points at the bottom of the map. Often an arrow is placed over which north (Grid or True) is being used on the map. This is simple if the Grid North of the map is the same as the True North and usually True North will be within 2° on either side of Grid North.

If the true north is located on the left hand side of grid north, then only the angle between grid and magnetic Norths (i.e. the GMA) is simply read off. If there is any confusion, just remember that conversion between a map and a compass concerns the angle between Grid North and Magnetic North. Looking at the map north points reference diagram should clearly show how to obtain this angle, whether there is any difference between True North and Grid North.

There is another complication however, because the angle between True North and Magnetic North (the magnetic declination), changes with <u>time</u> and is different in <u>different localities</u>.

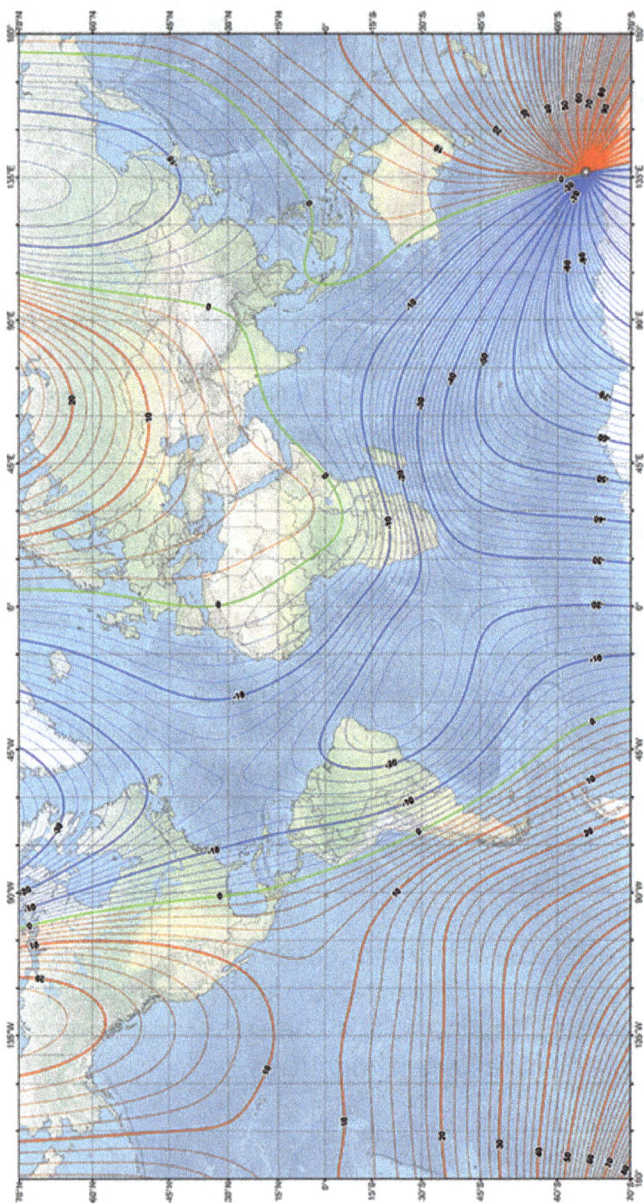

Figure 6.6: Worldwide Declinations. Easterly (positive) declinations are given in RED, westerly (negative declinations) are given in BLUE (Photo: NOOA)

For example, if the north points are positioned as in Figure 6.5 which shows a variation of Magnetic North easterly by 0.1^0 every four years, then the grid magnetic angle to be used when making conversions would be:

Total magnetic declination for 2016 (not 2000 as shown in Figure 6.5) =

(10^0 declination for the year 2000)   +   0.1° x 16 (i.e. the 16 years from year 2000)

= approximately 12°

So the total difference between Grid North and Magnetic North (i.e.the GMA)

= 2° + 12^0 = 14°

Now that the local declination for that specific locality has been calculated, the bearing on the map grid can then be converted to a magnetic bearing. Often, if the difference between Grid North and True North is very small, then grid north is taken as approximately equal to True North and only it is used in further calculations to avoid confusion.

Now keeping this calculation simple, the difference in angle between Grid North and Magnetic North must be added or subtracted depending upon the locality of the map in relationship to the lines of magnetic field (see Figure 6.6). If the locality has an easterly declination with Magnetic North to the right of Grid (True) North, the declination is subtracted from the grid bearing to get the magnetic bearing. For example, if the declination is 14° east and the bearing measured by a protractor on the map is 180^0 (i.e. south), then the compass would have to be set to 180^0 - 14° = 164° to actually walk due south.

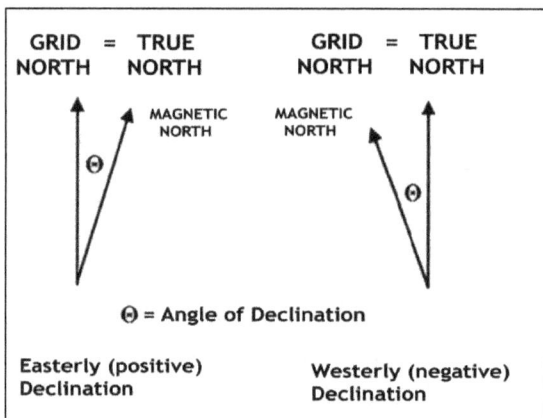

GRID = TRUE
NORTH   NORTH

MAGNETIC
NORTH

GRID = TRUE
NORTH   NORTH

MAGNETIC
NORTH

Θ

Θ

Θ = Angle of Declination

Easterly (positive)
Declination

Westerly (negative)
Declination

Figure 6.7: The relationships between true and magnetic norths

If the mapped region had a westerly declination, then the declination would be added to the grid bearing. For example if the declination was 4° west, and the grid bearing was again south, then the actual compass bearing would be 180^0 + 4° = 184°. So to avoid confusion, one must know the declination of the map locality in degrees and whether it is east or west (see Figure 6.6) and learn that with an EASTERLY declination one can remember:

G.M.S.

GRID (to) MAGNETIC SUBTRACT (declination)

but with a <u>westerly</u> declination it is <u>added</u>

However, whatever the declination is at one's locality, the rules given above are <u>reversed</u> when plotting a course <u>onto</u> a map (i.e. grid bearing) from a compass (magnetic) bearing. For example, if a compass bearing to a distant feature, such as a hill, is 078^0 magnetic (note the use of 3 digits here!) and the magnetic deviation is <u>easterly</u> at $14^{0,}$ then this variation is <u>added</u> i.e. $078^0 + 14^0 = 092^0$. This would be the angle that could be drawn from the observation locality on the map to the hill.

Such an operation is usually done when one wishes to find their current location by taking **back-bearings.** This is a method using **triangulation** which uses back-bearings from at least three prominent surface features (mainly hills and mountains) which are easily identified on the map. The back-bearings from each of these features are drawn from them onto the map and the intersection gives the current location.

6.3 More about Contour Lines

Contour lines join places of equal height, so their shape is representative of slopes and landforms which are useful when planning a traverse across the country. The closer the contour lines are together, the steeper the slope and, knowing the contour interval and the distance over several contours, one can determine the gradient. This is done by calculating the difference in height from the difference in the values of the contour lines and then dividing this by the horizontal distance between them. This will give some idea of the degree of difficulty in walking that distance and the time that it might take. Some of the most common landforms shown by contours are given in Figure 6.8.

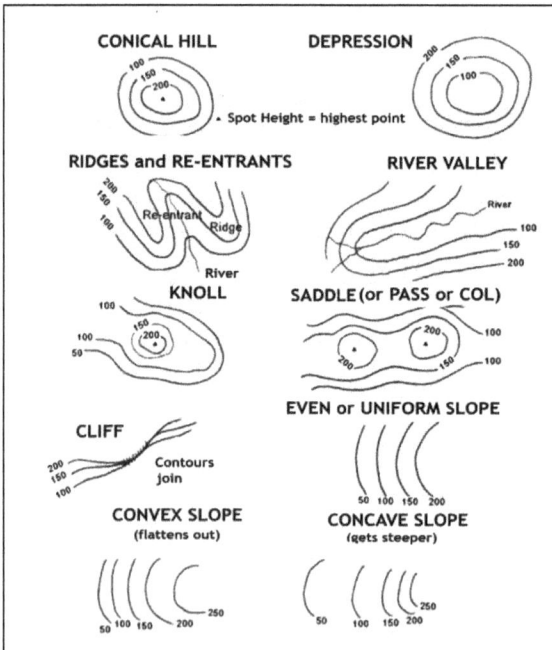

Figure 6.8: Diagram showing landforms represented by contour lines

6.4 Drawing a Cross-Section

Geologists are interested in what is going on below the ground so they may wish to draw a cross-section through the Earth along a particular line across the topographic map. For example, consider the following traverse A to B across a landform known as a saddle:

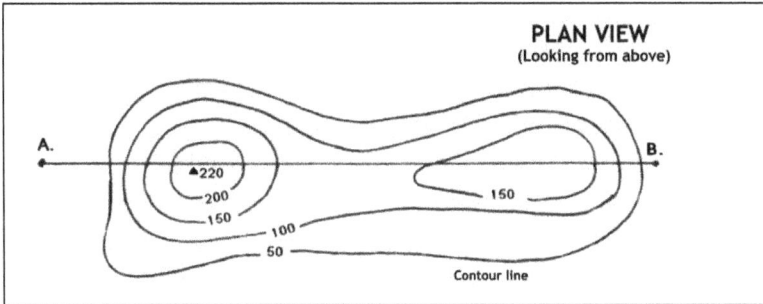

Figure 6.9: A topographical plan view of a feature called a saddle

The steps that a geologist would take in constructing a cross-section would be to:

1. Place the edge of a piece of paper along the A-B line on the plan view (map) and mark off along that edge the positions of A, B and each contour line and its height.

2. Decide on a vertical scale. This will probably not be the same as the horizontal scale because most lengths A-B would be in kilometres and depth might only be a few hundred metres. Usually this may be from the highest point (220 m in this case) down to sea level or a level that would give a good cross-

section. Using different vertical and horizontals scales introduces the use of **vertical exaggeration** in construction. This is the stretch factor that is made to give a cross-section of reasonable thickness, but it does cause later problems when sub-surface angles are drawn to scale because they have to be exaggerated also. In this example the V.E. will be = 2.

3. Draw up a cross-section box with horizontal length = AB to scale (here it is 1 cm = 100 metres) and vertical height using the vertical scale (say 1cm = 50 metres) from about 300 m above sea level down to sea-level;

4. Place the marked paper edge along the ab line of this box and transfer its data to the top (some use the bottom) of the box along the newly drawn line AB. Draw faint construction lines from each position marked (A,B, and contour heights) down to the appropriate height marked to scale in the box. This gives the rough profile.

5. Join up the dots (red in this example) with curves of best fit as a reasonable way that a land surface would look (no medical chart spikes here!). Erase any construction lines.

The newly drawn profile is ready for the sub-surface geological data. This is a very tiresome construction and there are computer modelling programs which do it much better.

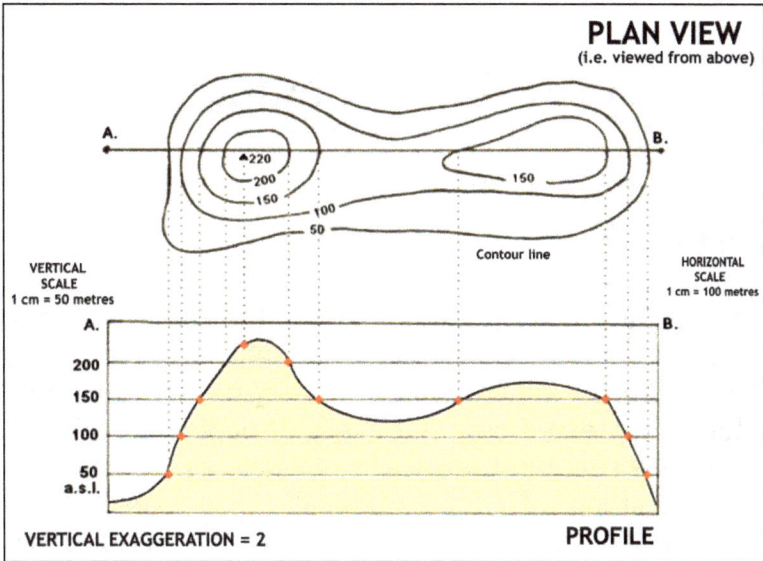

PLAN VIEW
(i.e. viewed from above)

A.

B.

▲220
200
150
100
50

Contour line

VERTICAL SCALE
1 cm = 50 metres

HORIZONTAL SCALE
1 cm = 100 metres

A.

B.

200
150
100
50
a.s.l.

VERTICAL EXAGGERATION = 2

PROFILE

Figure 6.10: The completed cross-section of the Saddle

Chapter 7: Geological Maps

7.1 Introduction

A geological map attempts to show the main rock and soil types, geological structures, mining details and other features which would be useful to the field geologist expressed on a flat sheet of paper. Usually the terrain, the real land surface is not flat, so the features of the standard topographical map are often superimposed onto the geological data. This often makes these maps difficult to read, so the different rock and soil types are often coloured and given codes corresponding to their type or age.

Geological maps are a convenient tool to use in the field to locate a starting point for new exploration, to find features, or deposits which have been already mapped. Like any other map, there are often degrees of uncertainty in the data shown on geological maps. Some large-scale areas, especially in difficult terrain, may have been mapped only from aerial or satellite surveys. These can be very good for general exploration, but their accuracy will not be as good as a small-scale area mapped by surveyors on the ground. It is important, then, to understand the main colours and symbols shown on geological maps and the features which they represent.

7.2 Mapping Horizontal Beds (symbol ╤)

Sedimentary rocks, volcanic ash falls, old lava flows and exposed volcanic sills are often in a horizontal orientation. If they cover a wide area, then walking across the surface will give nothing but the uniform geology of the top layer. One must get below the surface by investigating vertical cross-sections such as in gullies, river channels, road and rail cuttings (<u>caution</u>). For a more detailed and professional study, the area may have to be

Figure 7.1: Student geologists examining cores at a drill core library of a large mine

selectively drilled and the drill cores examined and measured and logged. This is done by the geologist carefully and accurately measuring each rock type with a measuring tape down the length of each drill core which may be several hundred metres long.

For example, consider the following surface with four sampling drill holes, A, B, C and D:

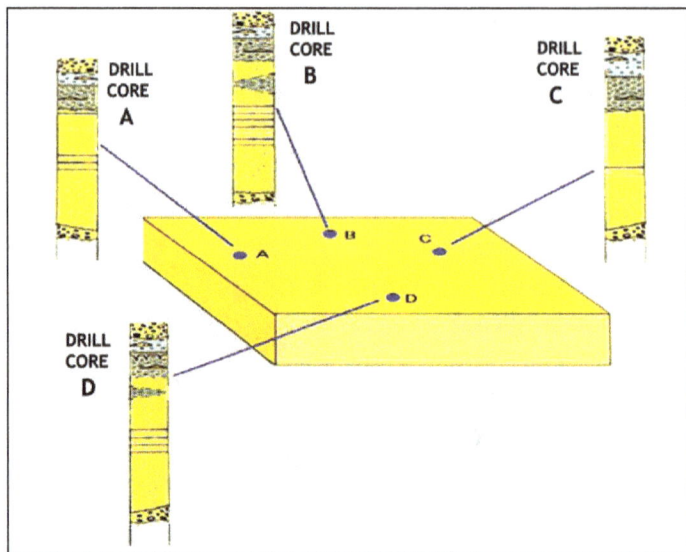

Figure 7.2: A Diagram showing several drill holes and the logs drawn as stratigraphic columns

Each **stratigraphic column** represents a visible cross-section made from the logging of each of the four cores. There is computer software designed to take a large number of logs and then present them as a **fence diagram** by matching and linking similar rock types, layers and structures together:

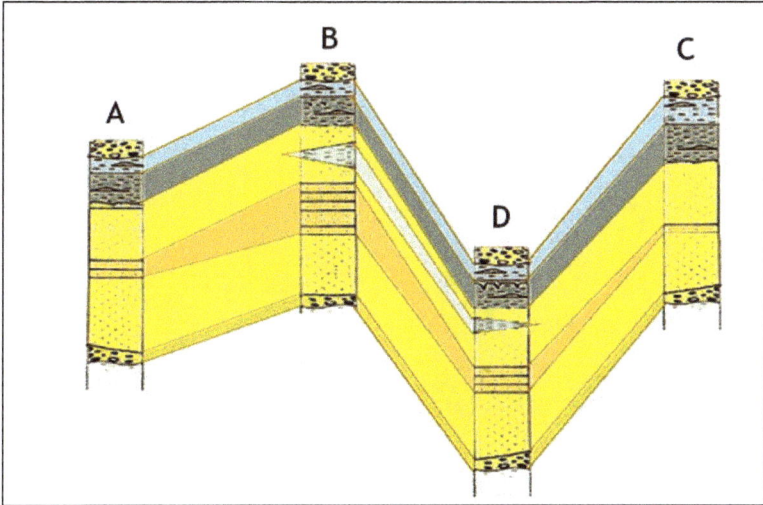

Figure 7.3: Fence diagram linking the strata of all four drill cores to give a three-dimensional view of sub-surface geology.

7.3 Mapping Tilted Beds (symbol —)

One common feature seen in landscapes, especially in road and rail cuttings, is sedimentary bedding and other normally horizontal features such as basalt lava flows, which have been tilted at an angle. This may have occurred because of upthrust by a nearby fault or fold, intrusion by a salt dome or igneous structure, or by a broad movement associated with the regional tectonic plate.

Figure 7.4: Tilted strata at Garganta del Diablo (the Devil's Throat) near Cafayete, Argentina (Photo: Matthew Scott)

There are several dimensions used in describing the orientation of any tilted bed or other planar geological surface. These are the:

- **Angle of dip** (or **dip**) which is the angle that the tilted beds are measured to the horizontal.

- **Dip direction** which is the geographical direction towards which the tilted beds are facing. This measurement and the angle of dip, are the two best measurements of the orientation of a sloping surface. <u>Dip direction is always at 90^0 to the strike</u> (see below) and

- **Strike** which is the geographical direction of the intersection of the beds with the surface, that is the way the outcrop of the beds appear to run across the surface of the land.

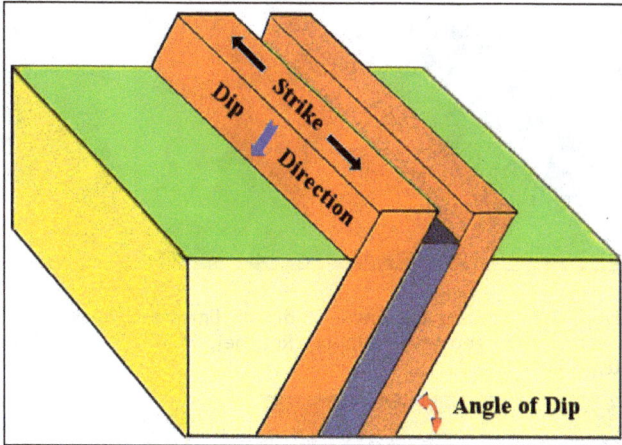

Figure 7.5: Diagram showing orientation dimensions of a tilted surface.

Measuring dip and dip direction can be done using a Brunton Pocket Transit (or simply called a Brunton Compass) which has been designed for geologists and surveyors. This is a versatile instrument for measuring the orientation of any planar surface.

Figure 7.6: Finding the angle of dip on limestone beds dipping towards the geologist (Upper Rouchel, New South Wales, Australia)

Figure 7.7: Sometimes the strike is not easy to see. (Kholo, Queensland, Australia)

The angle of dip is measured by firstly finding a satisfactory flat bedding surface. This is not often easy in the field due to weathering and fracturing of the rock. In addition, fracturing of the rock often also produces clean, flat fracture surfaces which may be at an oblique angle to the true direction of dip. Measuring on these flat fracture surfaces will give an **apparent dip** which may be wrongly taken as the true dip. An apparent dip will always be a smaller angle than the true dip.

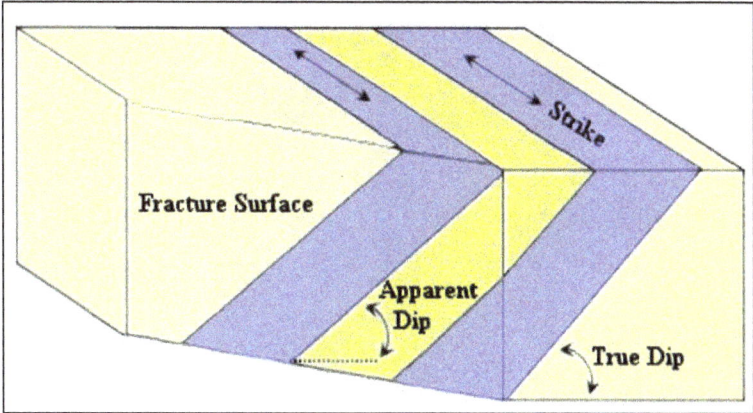

Figure 7.8: Diagram showing the misleading nature of apparent dip

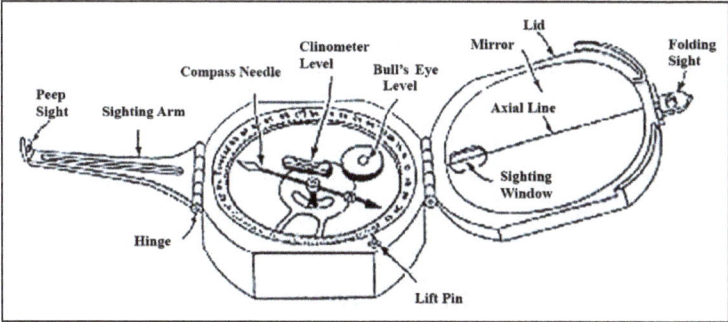

Figure 7.9: The main parts of a Brunton Pocket Transit

To measure the angle of dip, the compass is placed on its side with the sighting arm facing down the dip slope, as shown below. The small lever on the back of the compass is then turned so that the spirit level (Clinometer Level) on the moveable arm inside the compass is in its horizontal position seen when the bubble in it comes up to the horizontal mark. The angle is then read off the centre of small scale (Clinometer Scale) on the moveable arm.

Figures 7.10 & 7.11: Measuring the angle of dip (top) and reading off the value of the clinometer scale, showing an angle of about 8 degrees from the horizontal (bottom).

The dip direction is measured by holding the compass horizontally out from the dipping plane and at right-angles to the strike of the surface, that is in the direction of the dip, with its long sighting arm down the direction to be measured. Checking that the large, round spirit level in the compass (Bull's Eye Level) shows that it is horizontal, the geographical bearing can be read off from the main magnetic compass needle. The reading is always given as three digits as degrees magnetic. Dip direction is far less confusing than a strike bearing, which can be given as two bearings at 180^0 to each other e.g. 000^0 – 180^0 for north-south.

Figure 7.12: The dip direction is that direction along the sighting arm (here towards the right)

Online Video: 7.1: Measuring angle of dip and dip direction
Go to https://www.youtube.com/watch?v=BrTekM0DRQk

CAUTION - SMARTPHONES!

There are many geological **smartphone Apps** which can be useful for GPS navigation, data logging, reference and also measurement of dip, dip direction and strike. However, caution must be exercised because the Earth's magnetic field fluctuates rapidly and smartphone 3 axis magnetometers can potentially measure background magnetic noise. Local ironstone deposits and power lines will also cause disruption as they do to the Brunton Compass and all smartphones create strong magnetic fields that must be corrected by computer algorithms. Geological compass apps should be tested against traditional magnetic compass measurements before going into the field.

For reference and maps, good offline apps are available as internet connection may not be possible in many remote areas. In addition, one must consider the battery drain of the device and ensure that a re-charger is available back at base at the end of each day. For extended exploration away from base, a solar-powered battery charger with a short re-charge time should be considered.

In drawing a cross-section of tilted beds, a depth of cross-section is chosen to scale, with or without an appropriate vertical exaggeration. This is when the vertical scale is stretched to give a thicker cross-section so some detail can be seen more clearly. Next, the positions of the bed boundaries and their dips are marked off onto a strip of paper and then transferred to the top (or bottom) of the profile box.

But what happens if the traverse A-B is not down the dip direction but inclined at a geographical bearing angle of (say) 20° to the **direction angle**? The dip then being drawn is the apparent dip not the true dip. Remember that the apparent dip is always less than the true dip.

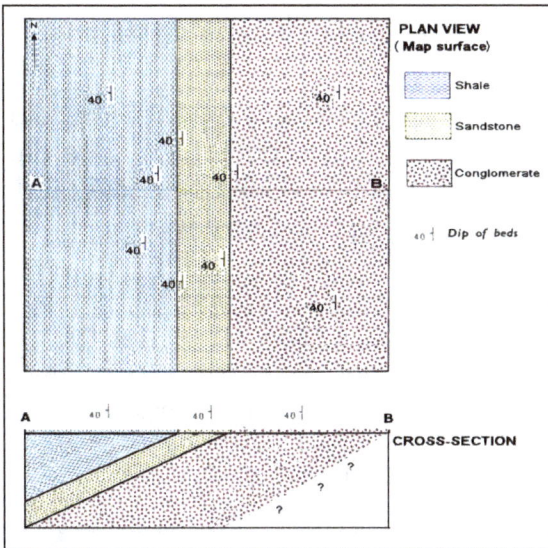

Figure 7.13: Diagram of plan view and a cross-section across A-B for beds dipping at 40° towards the west (dip direction) and no vertical exaggeration.

However, the next diagram represents a traverse A-B which is at an angle of 20⁰ to the dip direction.

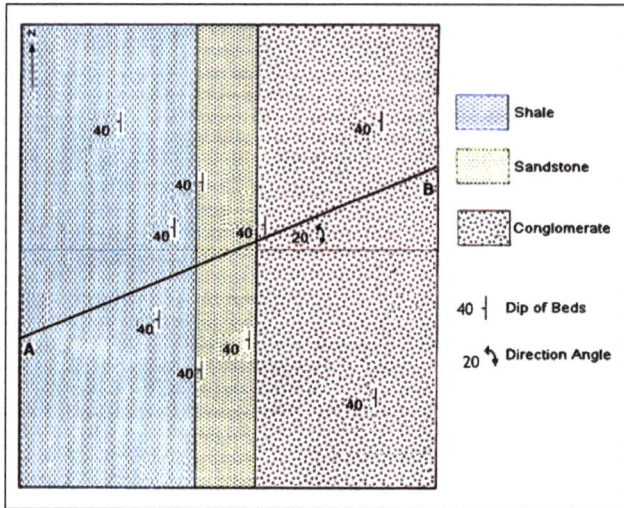

Figure 7.14: a traverse oblique to the dip direction

To draw an accurate cross-section with the dip shown along this section one must convert the true dip to the apparent dip. This is can be done using several different methods, often involving the calculation:

Tan AD = Tan TD x Cos DA

Where AD = Apparent Dip Angle;
TD = True Dip Angle; and
DA = Direction Angle.

So in this example:

$$\text{Tan AD} = \text{Tan } 40^0 \times \text{Cos } 20^0$$

$$= 0.8391 \times 0.9397$$

$$= 0.7885$$

Giving an angle of about 38^0 which is now drawn in as the dip of the beds:

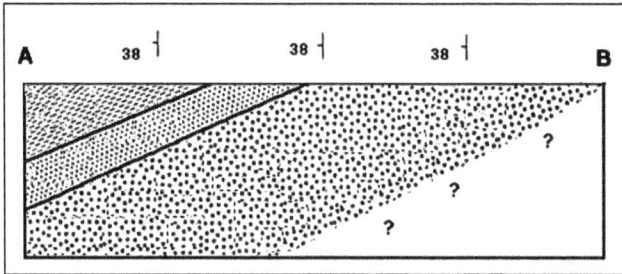

Figure 7.15: Apparent dip angles draw for cross-section A-B

As a guide for quick reference the following table (Table 7.1) can be used to obtain apparent dip from the true dip and the direction angle:

True Dip	Direction angle - angle between True Dip and Traverse															
	10°	15°	20°	25°	30°	35°	40°	45°	50°	55°	60°	65°	70°	75°	80°	85°
	Apparent dip															
10°	10°	10°	9°	9°	9°	8°	8°	7°	6°	6°	5°	4°	3°	3°	2°	1°
15°	15°	14°	14°	14°	13°	12°	12°	10°	10°	9°	8°	6°	5°	4°	3°	1°
20°	20°	19°	19°	18°	18°	17°	16°	14°	13°	12°	10°	9°	7°	5°	4°	2°
25°	25°	24°	24°	23°	22°	21°	20°	18°	17°	15°	13°	11°	9°	7°	5°	2°
30°	30°	29°	28°	28°	27°	25°	24°	22°	20°	18°	16°	14°	11°	9°	6°	3°
35°	35°	34°	33°	32°	31°	30°	28°	26°	24°	22°	19°	16°	13°	10°	7°	4°
40°	40°	39°	38°	37°	36°	35°	33°	31°	28°	26°	23°	20°	16°	12°	8°	4°
45°	45°	44°	43°	42°	41°	39°	37°	35°	33°	30°	27°	23°	19°	15°	10°	5°
50°	50°	49°	48°	47°	46°	44°	42°	40°	37°	34°	31°	27°	22°	17°	12°	6°
55°	55°	54°	53°	52°	51°	49°	48°	45°	43°	39°	36°	31°	26°	20°	14°	7°
60°	60°	59°	58°	58°	56°	55°	53°	51°	48°	45°	41°	36°	30°	24°	17°	9°
65°	65°	64°	64°	63°	62°	60°	59°	57°	54°	51°	46°	42°	36°	29°	20°	11°
70°	70°	69°	69°	69°	68°	67°	65°	63°	60°	58°	54°	49°	43°	35°	25°	13°
75°	75°	74°	74°	74°	73°	72°	71°	69°	67°	65°	62°	58°	52°	44°	33°	18°
80°	80°	80°	79°	79°	78°	78°	77°	76°	75°	73°	71°	67°	63°	56°	45°	26°
85°	85°	85°	85°	84°	84°	84°	83°	83°	82°	81°	80°	78°	76°	71°	63°	45°

Table 7.1: Giving apparent dips from true dip and direction angle. For example, If the true dip is 30° but the traverse is at 20° from the dip direction, then the apparent dip would be 28°

Sometimes the cross-section may be over a very long surface distance (in kilometres) but only have a depth of a few hundred metres. This would give a very long but extremely thin cross-section. To give a better representation of the sub-surface, the vertical scale may be exaggerated to make the cross-section readable.

A vertical exaggeration (V.E.) adds an extra complexity to drawing a cross-section because the angle of dip on the map will not translate accurately to the cross-section. The angle of the dipping beds must also be exaggerated on the cross-section to match. Recall that:

VERTICAL EXAGGERATION (V.E.) = HORIZONTAL SCALE
                                 VERTICAL SCALE

Taking a simple example, if there was no V.E. for a dip of 45°, then the horizontal distance would be equal to the vertical distance.

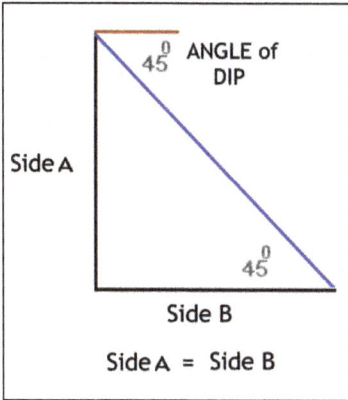

Figure 7.16: Dip of 45° with no vertical exaggeration i.e. Side A = Side B

However, if there is a V. E. of 2 is applied, then the side **A** would have to be twice its length and so the angle of dip would be steeper (about 63^0).

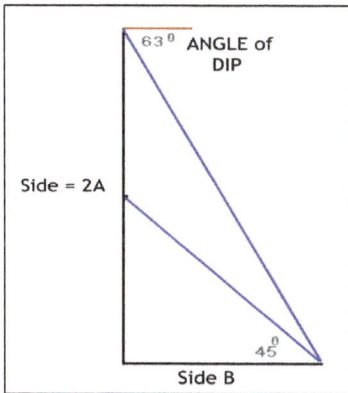

Figure 7.17: The exaggerated dip with a V.E. of 2 i.e. Side 2A = 2 x Side B

The new angle produced by vertically exaggerating the cross-section can be found from:

Tangent (or tan) α = **VE (Tan TD)**

Where α is the new exaggerated Angle of Dip;
VE is the Vertical Exaggeration (e.g. 2); and
TD is the True Dip (e.g. 45^0)

Or in this case: **Tan α** = **2 x tan 45°**

= 2 x 1 (since Tan 45° = 1)

So from tangent tables, the angle with a tangent of 2.0000 is approximately 63^0. This would then be the new angle of dip to be drawn on the cross-section with a V.E. of 2 to represent the true angle of dip of 45^0.

Table of Tan (angle)

Angle	tan(a)	Angle	tan(a)	Angle	tan(a)	Angle	tan(a)
0.0	0.00	25.0	.4663	46.0	1.0355	71.0	2.9042
1.0	.0175	26.0	.4877	47.0	1.0724	72.0	3.0777
2.0	.0349	27.0	.5095	48.0	1.1106	73.0	3.2709
3.0	.0524	28.0	.5317	49.0	1.1504	74.0	3.4874
4.0	.0699	29.0	.5543	50.0	1.1918	75.0	3.7321
5.0	.0875	30.0	.5773	51.0	1.2349	76.0	4.0108
6.0	.1051	31.0	.6009	52.0	1.2799	77.0	4.3315
7.0	.1228	32.0	.6249	53.0	1.3270	78.0	4.7046
8.0	.1405	33.0	.6494	54.0	1.3764	79.0	5.1446
9.0	.1584	34.0	.6745	55.0	1.4281	80.0	5.6713
10.0	.1763	35.0	.7002	56.0	1.4826	81.0	6.3138
11.0	.1944	36.0	.7265	57.0	1.5399	82.0	7.1154
12.0	.2126	37.0	.7535	58.0	1.6003	83.0	8.1443
13.0	.2309	38.0	.7813	59.0	1.6643	84.0	9.5144
14.0	.2493	39.0	.8098	60.0	1.7321	85.0	11.430
15.0	.2679	40.0	.8391	61.0	1.8040	86.0	14.301
16.0	.2867	41.0	.8693	62.0	1.8907	87.0	19.081
17.0	.3057	42.0	.9004	63.0	1.9626	88.0	28.636
18.0	.3249	43.0	.9325	64.0	2.0503	89.0	57.290
19.0	.3443	44.0	.9657	65.0	2.1445	90.0	Infinite
20.0	.3640	45.0	1.000	66.0	2.2460		
21.0	.3839			67.0	2.3559		
22.0	.4040			68.0	2.4751		
23.0	.4245			69.0	2.6051		
24.0	.4452			70.0	2.7475		

Table of Cos (angle)

Angle	cos(a)	Angle	cos(a)	Angle	cos(a)	Angle	cos(a)
0.0	1.00	25.0	.9063	46.0	.6947	71.0	.3256
1.0	.9998	26.0	.8988	47.0	.6820	72.0	.3090
2.0	.9994	27.0	.8910	48.0	.6691	73.0	.2924
3.0	.9986	28.0	.8829	49.0	.6561	74.0	.2756
4.0	.9976	29.0	.8746	50.0	.6428	75.0	.2588
5.0	.9962	30.0	.8660	51.0	.6293	76.0	.2419
6.0	.9945	31.0	.8571	52.0	.6157	77.0	.2249
7.0	.9926	32.0	.8480	53.0	.6018	78.0	.2079
8.0	.9903	33.0	.8387	54.0	.5878	79.0	.1908
9.0	.9877	34.0	.8290	55.0	.5736	80.0	.1736
10.0	.9848	35.0	.8191	56.0	.5592	81.0	.1564
11.0	.9816	36.0	.8090	57.0	.5446	82.0	.1392
12.0	.9781	37.0	.7986	58.0	.5299	83.0	.1219
13.0	.9744	38.0	.7880	59.0	.5150	84.0	.1045
14.0	.9703	39.0	.7772	60.0	.5000	85.0	.0872
15.0	.9659	40.0	.7660	61.0	.4848	86.0	.0698
16.0	.9613	41.0	.7547	62.0	.4695	87.0	.0523
17.0	.9563	42.0	.7431	63.0	.4540	88.0	.0349
18.0	.9511	43.0	.7314	64.0	.4384	89.0	.0174
19.0	.9455	44.0	.7193	65.0	.4226	90.0	0.0
20.0	.9397	45.0	.7071	66.0	.4067		
21.0	.9336			67.0	.3907		
22.0	.9272			68.0	.3746		
23.0	.9205			69.0	.3584		
24.0	.9135			70.0	.3420		

Table 7.2 Tangent and Cosines

7.4 Mapping Folded Beds (symbols ⟨⇌⟩ or ⇥|⟨)

When pressure is applied to strata, it can fold upwards as **anticlines** or downwards as **synclines,** or sometimes as both in a series of complex folds. If forces from other directions are also applied, these folds can also be tilted in another direction as **plunging folds.**

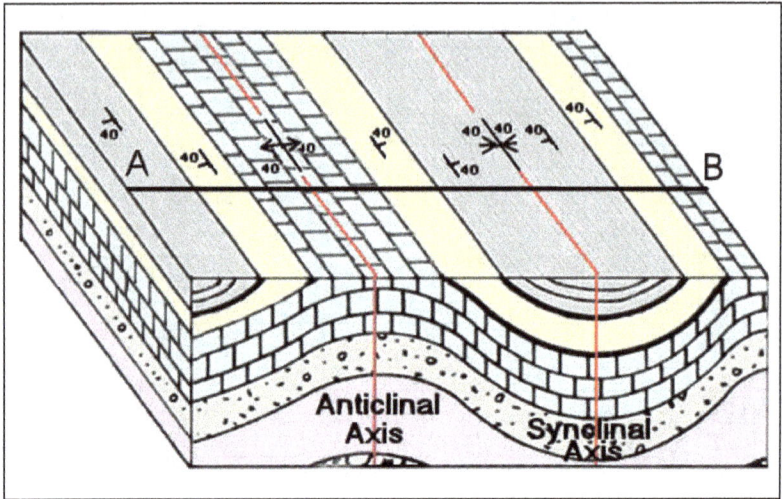

Figure 7.18: Diagram in plan and sectional view of an anticline and syncline

The folds shown in the diagram above are symmetrical folds, having the same dip angles but opposite directions on either sides of their axes. Asymmetrical folds have different dip angles on each side of their axes and give a similar cross-section to the one given above except that the dip angles will be different.

When constructing a cross-section from data obtained from a traverse across the dip directions i.e. at 90^0 to the strike of the beds and the fold axes, the following constructions are made:

1. Draw a rectangular box for the cross-section choosing an appropriate scale (in this example there is no V.E)

2. Use a paper edge to mark off the boundaries of the beds and the fold axes of each fold

3. In the box draw in faint vertical construction lines for each of the fold axes

4. Having marked off the position of bed boundaries, use a protractor to make faint construction lines for all of the dipping bed boundaries (scale angles if a V.E. is used)

5. Draw in the curves of the folds along the construction line using a line of best fit;

6. Complete the cross-section by using the appropriate rock symbols and colours.

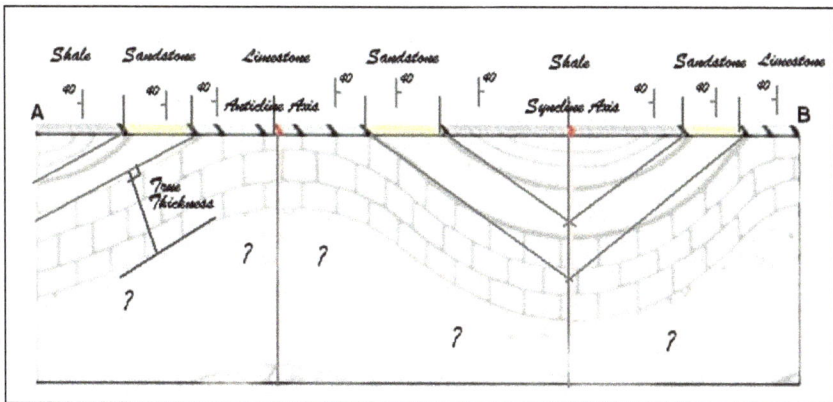

Figure 7.19: Constructed cross-section showing rough working lines

In this diagram, the thickness of the limestone bed would have to be estimated using the surface width of approximately half the length across the fold axis and the angle of dip (here 40⁰) would be calculated using trigonometry:

105

Thickness of bed = Sine (Angle of Dip) x Outcrop Width

Table of sin (angle)							
Angle	sin (a)	Angle	sin (a)	Angle	sin (a)	Angle	sin (a)
0.0	0.0	25.0	.4226	46.0	.7193	71.0	.9455
1.0	.0174	26.0	.4384	47.0	.7314	72.0	.9511
2.0	.0349	27.0	.4540	48.0	.7431	73.0	.9563
3.0	.0523	28.0	.4695	49.0	.7547	74.0	.9613
4.0	.0698	29.0	.4848	50.0	.7660	75.0	.9659
5.0	.0872	30.0	.5000	51.0	.7772	76.0	.9703
6.0	.1045	31.0	.5150	52.0	.7880	77.0	.9744
7.0	.1219	32.0	.5299	53.0	.7986	78.0	.9781
8.0	.1392	33.0	.5446	54.0	.8090	79.0	.9816
9.0	.1564	34.0	.5592	55.0	.8191	80.0	.9848
10.0	.1736	35.0	.5736	56.0	.8290	81.0	.9877
11.0	.1908	36.0	.5878	57.0	.8387	82.0	.9903
12.0	.2079	37.0	.6018	58.0	.8480	83.0	.9926
13.0	.2249	38.0	.6157	59.0	.8571	84.0	.9945
14.0	.2419	39.0	.6293	60.0	.8660	85.0	.9962
15.0	.2588	40.0	.6428	61.0	.8746	86.0	.9976
16.0	.2756	41.0	.6561	62.0	.8829	87.0	.9986
17.0	.2924	42.0	.6691	63.0	.8910	88.0	.9994
18.0	.3090	43.0	.6820	64.0	.8988	89.0	.9998
19.0	.3256	44.0	.6947	65.0	.9063	90.0	1.00
20.0	.3420	45.0	.7071	66.0	.9135		
21.0	.3584			67.0	.9205		
22.0	.3746			68.0	.9272		
23.0	.3907			69.0	.9336		
24.0	.4067			70.0	.9397		

Table 7.3: Sines for given angles

Other data for the rocks below the limestone would have to be obtained from drill cores or other sources such as further mapping beyond points A and B.

7.5 Mapping Plunging Folds (symbols ⊕ ⋇ - vertical arrow shows the direction of the plunge)

Plunging folds are those folded beds which have undergone a second event of pressure which has then also tilted them in another direction (called the **plunge**). In plan view (from the top), these plunging folds are seen as curves rather than parallel beds. Across the folds (section A-B below), the cross-section is similar to that of any anticline-syncline section (see Figure 7.19. Along the direction of the plunge (section C-D below), the cross-section simply shows parallel dipping along the direction of plunge as in Figure 7.15.

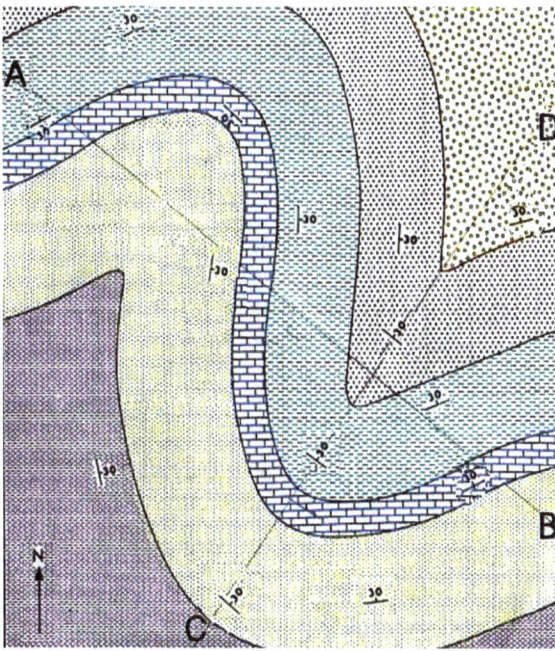

Figure 7.20: Showing a plan view of plunging folds

7.6 Mapping - Faults (symbols F ———— F)

If sudden forces are applied to the land, the rock layers may suddenly break and move, causing large faults. Faults are seen on the surface as sudden changes in lithology (rock type), a displacement of a linear feature such as a river, or a sudden long line of cliffs as a **fault scarp.** Faults are often very prominent in aerial and satellite photographs as long lines. Sometimes the fault may have a covering of soil and/or vegetation and may only be seen if there has been lateral displacement of a stream or a rocky outcrop. A fault may go unnoticed if it has occurred within a bed or parallel to a boundary. They become obvious when they cut obliquely across strata or when they are seen in profile within a rock face or cutting. Moreover, if the fault is vertical and there is no corresponding bedding on either side of it, called **marker beds**, then it is difficult to measure its vertical displacement or its **throw.**

The Great Glenn Fault of Scotland is easily seen as it divides the south from the north along Loch Ness:

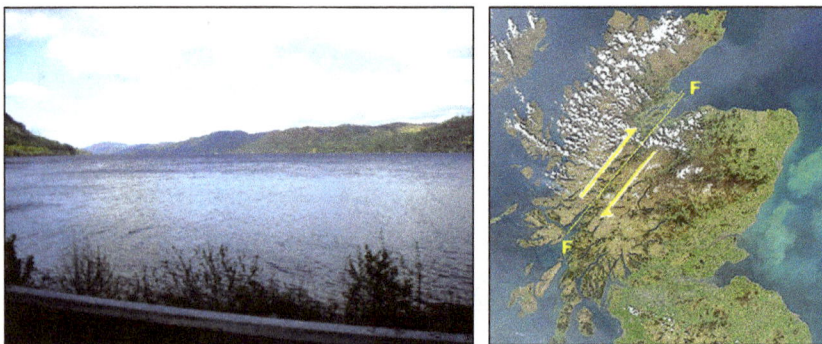

Figures 7.21 & 7.22: The long straight line of the hills on the far side of Loch Ness, Scotland marks the Great Glenn Fault (left photo: modified from NASA).

For mapping a fault, the construction sequence would be:

1. Draw a rectangular box for the cross-section choosing an appropriate scale (in this example there is no V.E)

2. Use a paper edge to mark off the boundaries of the beds and the fault position

3. Draw in a heavy line for the fault dipping at 80^0 W

4. Using the position of the boundaries of the beds, draw in the tops of each bed

5. Calculate the thickness of the sandstone bed from its outcrop length and dip using the previous formula (thickness of bed = sine [angle of dip] x outcrop width) and draw in the base of the sandstone on the right hand side of the fault

6. Extend that thickness to the left hand side of the fault by measuring it at right angles from the top of the sandstone bed

7. Draw a dotted line from the far right hand side corner representing an uncertain base of the red conglomerate

8. Add rock texture and colour.

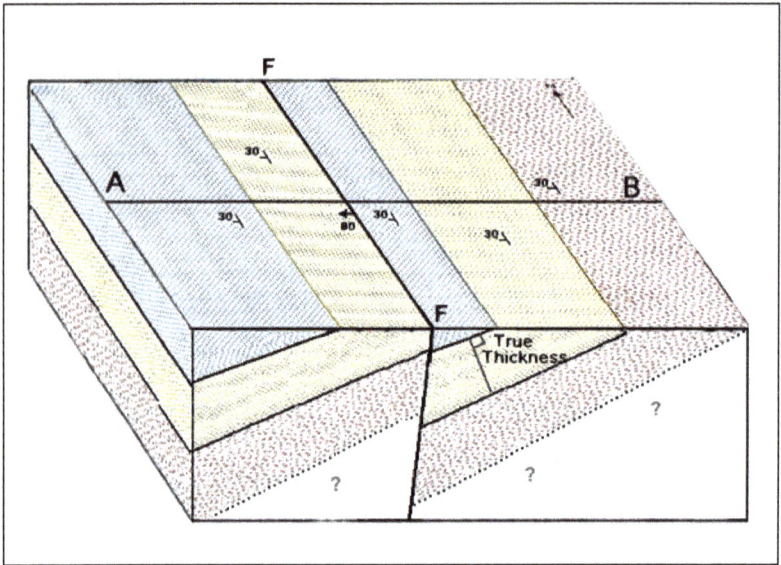

Figure 7.23: Diagram showing plan view and cross-section of a normal fault which is dipping to the west at 80°.

The **throw** or vertical displacement, and the **heave**, the horizontal displacement, can be measured from the scale of the map.

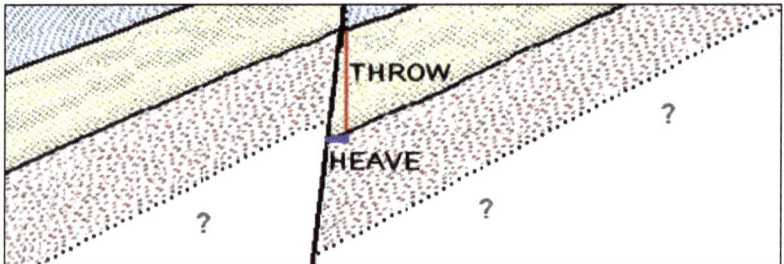

Figure 7.24: Diagram showing the cross-section for the dipping fault

7.7 Mapping Intrusions and Metamorhism

Igneous intrusions are relatively easy to locate because of the way they outcrop as batholiths and smaller stocks or volcanic necks, which usually stand out as tall, near-vertically sided peaks. They may also be completely eroded to a flat surface but may then be detected by the tell-tale piles of rounded tors on their surface or with extensive eroded, round, flat valleys containing a few outcrops of igneous rock. Batholiths and stocks in plan view appear as almost circular or rounded structures.

Figure 7.25: A quarry within a small stock of granite. Notice the rounded curve of the weathered top of the stock

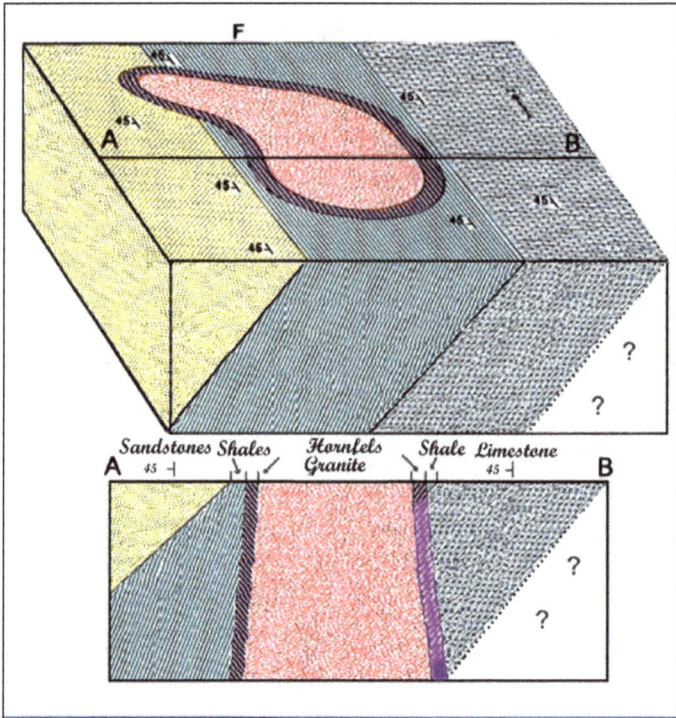

Figure 7.26: A diagram showing side and plan views of a batholith

Construction of the above cross-section would involve the following steps:

1. Draw a rectangular box for the cross-section choosing an appropriate scale (in this example there is no V.E)

2. Use a paper edge to mark off the boundaries of the beds and the edges of the batholith and the **metamorphic aureole** next to it (left and right)

3. Assume that the batholith has intruded upwards and so draw approximate sides enlarging down to the base. This is really unknown and it would require drill core sampling around the edges of the batholith to establish its true shape

4. Follow this shape with an approximate width for the metamorphic aureole – assuming that its width would be uniform down the sides of the batholith

5. Draw in the base of the sandstone dipping at 45^0 to the west

6. Draw a parallel bottom boundary for the shale on the right hand side of the batholith but only up to the hornfels (metamorphosed shale) of the aureole

7. Draw a line of uncertainty from the top right hand corner of the box for the unknown base of the limestone and

8. Label the right hand aureole below the hornfels as marble indicating the metamorphosis of the limestone.

Dykes and sills are thin, tabular structures formed as molten igneous rock is forced up along joints almost vertically through strata as dykes, or squeezed between rock layers and parallel to the bedding plane as sills. Exposed on the surface, sills can resemble lava flows, forming wide, very flat tops to hills as plateaux, but they usually have larger crystals and lack the gas vesicles and other features of flows. Dykes, as their name suggests are discordant, that is have cut through the country rock and now stand out as raised rock walls or as long vertical clefts left as the igneous rock weathers away within the surrounding hardened rock.

Figure 7.27: The Breadknife - an exposed dyke in the Warrumbungle National Park. New South Wales, Australia

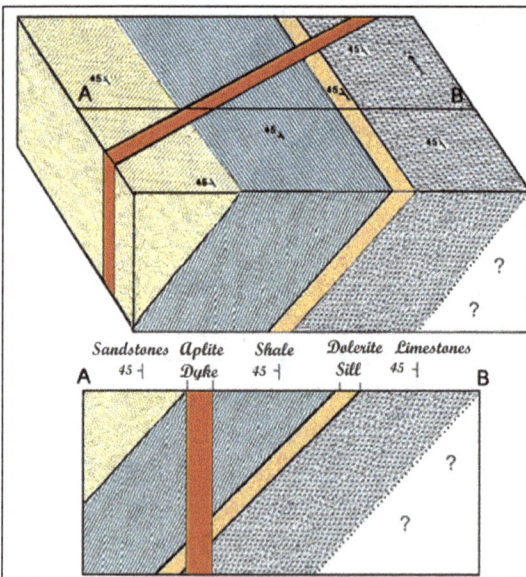

Figure 7.28: Plan view and cross-section of a vertical dyke and a sill dipping at 45⁰ parallel to the strata which it has intruded

Construction of a cross-section for the map (Figure 7.28 above) involves the following sequence:

1. Draw a rectangular box for the cross-section choosing an appropriate scale (in this example there is no V.E);

2. Use a paper edge to mark off the boundaries of the beds and the edges of the dyke;

3. Draw vertical lines straight down for the dyke as it is assumed that they are a vertical intrusion (unless marked with a dip) which cuts across (are discordant to) the other beds; and

4. Draw in the bottom boundaries to the dipping beds including the uncertainty of the base of the limestone.

Note that the dolerite sill is concordant, that is it is parallel to the bedding of the other rocks as it has penetrated between the sedimentary strata. The dyke on the other hand, is the youngest of all the rocks in the cross-section as it has cut across and intruded all of the existing rocks in the sequence. Aplite is a very fine felsic (quartz-rich) igneous rock whereas the older dolerite is a mafic igneous rock, suggesting two different periods of igneous activity.

7.8 Mapping Unconformities

Sedimentary deposition and the laying down of rocks within a basin usually does not continue without interruption. Strata can be tilted, folded and faulted, intruded and the surface is always weathered and eroded over time. Often there may be large gaps in time between the different episodes of sedimentation or rock disruption. The period of time in which there is no sedimentation but only erosion, is a called a geological **hiatus**. Later, there may be a new period of sediment deposition, volcanic activity or structural upheaval. When the two (or more) different episodes of rock formation are found within a sequence of rocks they are called an **unconformity** because one sequence of rocks does not conform to the original sequence of rock deposition.

In the diagram below (Figure 7.30), the cross-section is identical to the last exercise except that there has been erosion in the east forming a basin into which new sediment has been deposited. When mapping along this traverse, these new boundaries have to be marked off as before but superimposed upon the older beds and structures below. Note that the new sediments tilt eastwards at the edge but become horizontal (symbol: +) in the centre of the basin.

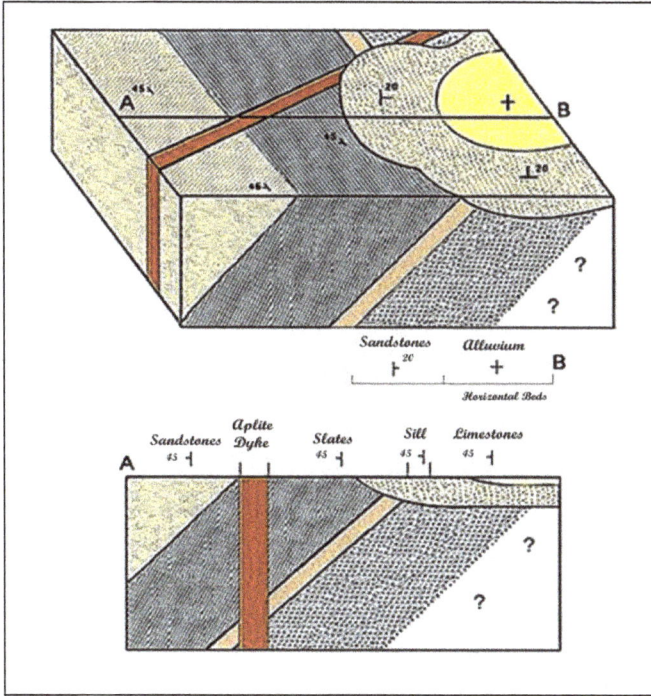

Figure 7.30: A diagram showing plan view and cross-section of an unconformity of new sedimentary beds (at right) over older beds

The term unconformity is a general term and there are specific types of unconformities. These include:

- **Disconformities** are breaks in the sedimentation between parallel layers of sedimentary rocks with little change in the angle of the bedding. This represents a period of erosion or non-deposition and there may be some slight features of erosion such as stream channels and soils at the boundary.

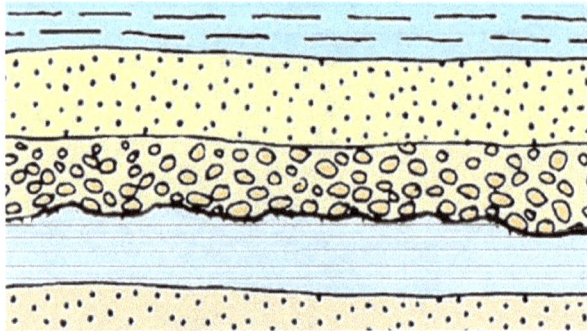

Figure 7.31: A disconformity

- **Paraconformities** are a type of disconformity in which the separation is a simple bedding plane with no obvious break in sedimentation. Whilst there has been a period where sedimentation has stopped and the beds were eroded, new sedimentation covered the old surface in the same orientation. These unconformities are often difficult to detect in the field.

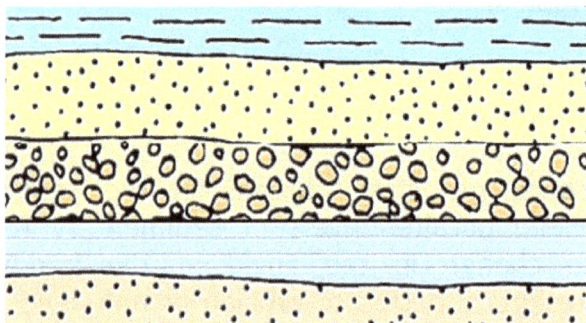

Figure 7.32: A paraconformity

- **Nonconformities** occur between sedimentary rocks and metamorphic or igneous rocks on which the new sediments have been deposited after the period of erosion.

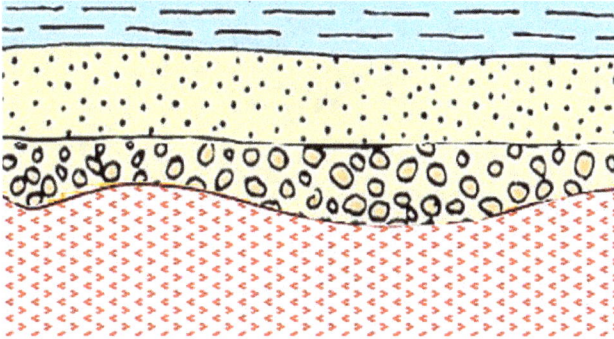
Figure 7.33: A nonconformity

- **Angular Unconformities** occur when horizontally parallel strata of sedimentary rock are deposited on tilted and eroded layers, producing an angular discordance with the overlying horizontal layers.

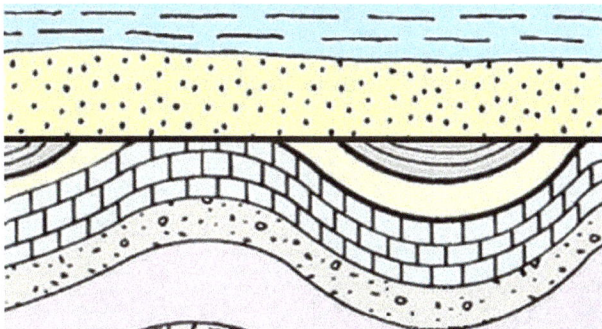
Figure 7.35: An angular unconformity

7.34 An angular unconformity, Capital Circuit, Canberra, Australia

7.9 Mapping Symbols and Colours

Most printed geological maps use conventional geological symbols and colours for rocks and sediment. As a quick reference to interpreting geological maps it is useful to have some knowledge of these symbols and colours.

The main mapping symbols include:

• General Reference and Surface Features:

TOPOGRAPHIC CONTOUR	——— *200* ———	WOODLAND	
BATHYMETRIC CONTOUR	—— *1000* ——	ORCHARD	
STREAM - PERENNIAL		SCRUB	
STREAM - INTERMITTENT		ROAD or STREET	———————
BRAIDED STREAM -PERENNIAL		PRIMARY ROUTE	———————
LAKE or POND		RAILROAD (one track)	─+─+─+─+─
INDUNATION AREA		PIPELINE	_ *PIPELINE* _
MARSH, WETLAND, SWAMP or BOG		POWER TRANSMISSION LINES	*POWER TRANSMISSION LINE*
MANGROVE AREA			
RICE FIELD			
CORAL REEF			

Table 7.4: Some map symbols of surface features

• Bedding:

KEY BED - CERTAIN	———————
KEY BED - APPROXIMATE	— — — — —
KEY BED APPORXIMATE, QUERIED	—?— — —?—
HORIZONTAL BEDDING	⊕
HORIZONTAL BEDDING from AERIAL PHOTOS	⋰
INCLINED BEDDING - strike & dip	⊥⁴⁰
OVERTURNED BEDDING	⁸⁵

Table 7.5: Geological bedding symbols

- ## Geological Structures:

DYKE - CERTAIN	————————	ANTICLINE - CERTAIN	
DYKE - APPROXIMATELY LOCATED	— — — — —	PLUNGING ANTICLINE - SHOWING PLUNGE	
KEY BED - APPROXIMATELY, LOCATED QUERIED	—?— — —?—	OVERTURNED ANTICLINE	
FAULT - CERTAIN	————————	SYNCLINE - CERTAIN	
FAULT - APPROXIMATELY LOCATED	— — — — —	OVERTURNED SYNCLINE	
FAULT - SHOWING DIRECTION and PLUNGE	65	MONOCLINE - CERTAIN	
NORMAL FAULT - CERTAIN			
STRIKE-SLIP FAULT - CERTAIN			
THRUST FAULT - CERTAIN			

Table 7.6: Some symbols for geological structures

- ## Fossil Types:

MACROFOSSILS		CRINOIDS	
TRACE FOSSILS		DIATOMS	
MARINE FOSSILS	M	ECHINOIDS	
PLANT REMAINS		FISH REMAINS	
LEAVES		FORAMS	
AMMONITES		GASTROPODS	
BELEMNITES		GRAPTOLITES	
BONES		INSECTS	
BRACHIOPODS		PELECYPODS	
BRYOZOA		STROMATOLITES	
CEPHALOPODS		TRILOBITES	
CORALS		VERTEBRATES	

Table 7.7: Symbols for fossil types within sedimentary rocks

• Resources:

SAND,GRAVEL,CLAY OR PLACER PIT	✕	GRADED AREA	(graded area symbol)
OPEN PIT, QUARRY or GLORY HOLE	⚒	DRILLING WELL	○
OPEN PIT MINE or QUARRY (surfacew view)	(symbol)	OIL WELL	●
SUBSURFACE WORKINGS (projected to the surface)	(symbol)	ABANDONED OIL WELL	(symbol)
STRIP MINE	(symbol)	CAPPED OIL WELL	(symbol)
VERTICAL MINE SHAFT	▱	GAS WELL	✿
INCLINED TUNNEL or ADIT	⟼	ABANDONED GAS WELL	(symbol)
TAILINGS POND (surface view)	(symbol)	HAZARDOUS WASTE SITE	▽

Table 7.8: Symbols for resources exploitation

• Tectonics - General:

ACTIVE SPREADING AXIS or MID-OCEANIC RIDGE	(symbol)
ACTIVE CONVERGENT PLATE BOUNDARY - CERTAIN	(symbol)
ACTIVE TRANSFORM FAULT - CERTAIN	(symbol)
ACTIVE TRANSFORM FAULT, RIGHT-LATERAL OFFSET - CERTAIN	(symbol)
CONTINENTAL SLOPE -CERTAIN	(symbol)
GUYOT	☼
SEAMOUNT - VOLCANIC	(symbol)
RIM of VOLCANIC CRATER - SHOWING the LOW POINT of the CRATER as a dot	(symbol)

Table 7.9: Symbols for tectonic structures

123

- Tectonics – Earthquakes:

EARTQUAKE EPICENTRES 7.5 or LARGER	◎
EARTHQUAKE EPICENTRE 7 to 7.49	●
EARTHQUAKE EPICENTRE 6.5 to 6.99	◉
EARTHQUAKE EPICENTRE 6 to 6.49	●
EARTHQUAKE EPICENTRE 5.5 to 5.99	○
EARTHQUAKE EPICENTRE 4 to 5.49	○
EARTHQUAKE EPICENTRE less than 4	○
FAULT SCARP - CERTAIN	⊢⊣⊣⊣⊣⊣

Table 7.10: Symbols for earthquake features

- Rock Symbols:

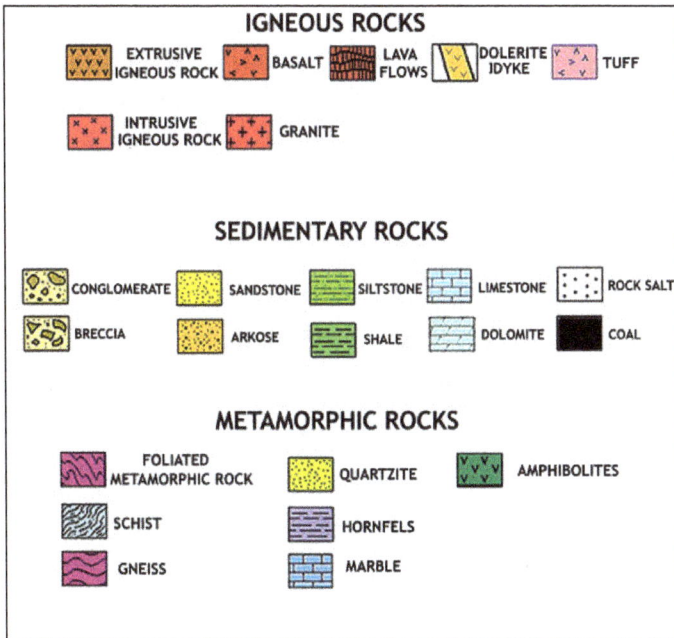

Table 7.11: Symbols for some common rock types

Colours are used, often in conjunction with rock symbols to indicate the lithology on the map. In general, the following standard colours are used:

- Igneous rocks tend to be oranges and reds

- Sedimentary rocks are greens and blues

- Metamorphic rocks are darker, murkier colours such as mauves, purples and the like and

125

- Alluvium deposits (unconsolidated sands and soils) are yellows and light browns.

The United States Geological Survey (USGS) has a set of standard colours for the various Geological Ages, with the colours getting darker for the earlier and often metamorphosed rocks. These colours appear to be in common use throughout the world on geological maps:

Table 7.12: Colour chart for stratigraphic rock units based on the USGS Reference Codes. Numbers refer to exact colour shadings used in computerized printing.

7.10 Geological History

One of the main reasons for studying an area, drawing a geological map and constructing a cross-section is to determine the events which happened in past geological. This reconstruction of geological events through time is called the **geological history** of the area.

Reconstruction of this type can be made using the Law of Uniformitarianism and the **Law of Superposition** as well as remembering how sedimentary, igneous and metamorphic rocks are formed. From these basic principles there are several main points to remember:

- Older rocks are found below younger rocks (the Law of Superposition).

- Younger structures such as faults, igneous intrusions, cut through older rocks.

- Intrusive igneous rocks are formed deep below the surface, so if they are exposed much uplift and erosion must have occurred.

- Sedimentary rocks are formed as horizontal layers from eroded, transported and deposited sediment, and that each rock type indicates a past environment.

ROCK TYPE	ANCIENT ENVIRONMENT
Conglomerate	pebbles from fast mountain rivers or storm beaches;
Sandstone	sand in medium-speed rivers or on coastlines of seas;
Shales	fine sediments deposited as muds in quiet waters of lakes and seas;
Limestones	coral reefs in warms, shallow, tropical seas;
Greywacke	rapidly deposited in deep ocean trenches;
Breccia	broken fragments quickly deposited by near volcanic by explosions or near ancient cliffs as landslides
Tuff	near volcanoes due to ash falling (often in seas or lakes)
Ignimbrites	welded ash formed as hot molten ash quickly falls;
Tillite	at the outwash end where glaciers melt to give streams of water;
Coal	vegetation near freshwater swamps or lakes.
Oil Deposits	Sea-floor conditions from dead marine organisms in muds

Table 7.13: Rock types and their original environments

The presence of certain fossils can also indicate the original environment:

FOSSIL TYPE	ANCIENT ENVIRONMENT
Plant Fossils	near land, especially near swamps, lakes and rivers.
Corals	warm shallow seas
Trilobites, Brachiopods, Molluscs	on relatively shallow ocean floors
Foraminifera,	exist in seas (and can be used to indicate temperature and
Vertebrates	are greatly varied in habitat but very indicative of conditions.

Table 7.14: Some fossil indicators of ancient environments

Even after a sedimentary rock sequence has been formed, there can be subsequent changes which add to the area's geological history. Intrusions of igneous rock can occur as a variety of different structures such as volcanoes, batholiths, dykes and sills. Faulting and folding can re-shape the strata and general uplift with weathering and erosion can occur in several different episodes to make the geology of an area very complicated. It is the task of the historical geologist to use the maps and cross-sections provided and considerable knowledge to unravel the geological history.

For example, consider a previous mapping sample:

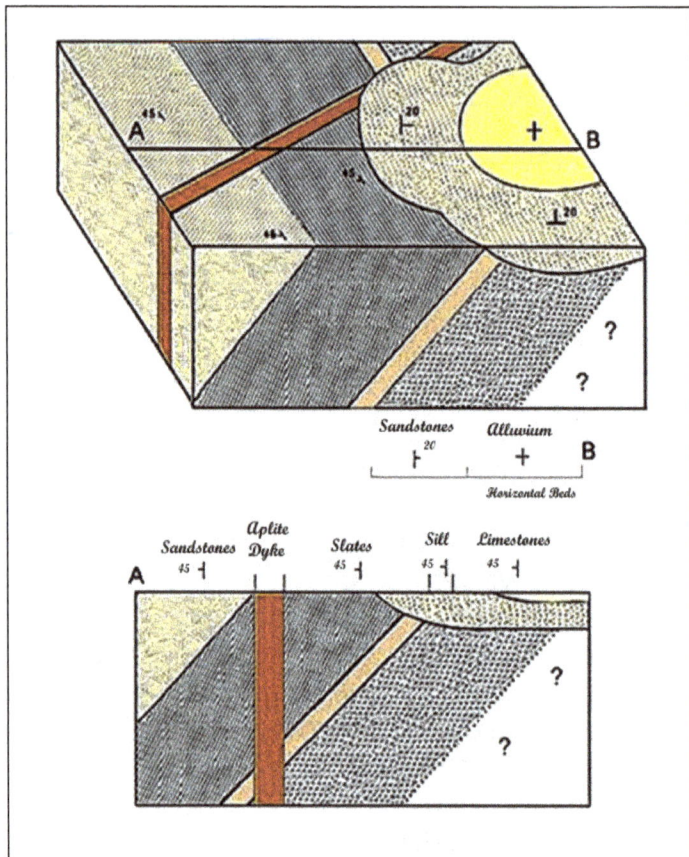

Figure 7.36: Sample of a map showing geological features

Beginning with the underline oldest underline events given first:

1.   The sedimentary rocks were deposited (in order) as:

   • Limestone (probably from marine corals so it is reasonable to assume that this is a marine sequence) then

- Shales (possibly marine) which may represent a quieter deposition of muds – perhaps the corals were inundated or the reef sank and then the sandstones formed from sand following a more violent period of wave action and rapid deposition.

2. Intrusion by the dolerite sill from some volcanic activity between the sedimentary strata.

3. Tilting of the sequence, perhaps by action associated with the volcanism.

4. Intrusion of the aplite dyke at some later stage so that it cut through all previously existing rocks.

5. Erosion in the eastern area of the site producing a shallow basin.

6. New sedimentation within this basin with the deposition of young sandstones and later younger, unconsolidated alluvium and finally

7. Erosion of the entire area to a flat surface. This later process is sometimes forgotten by student geologists!

These are the methods which the field geologist employs to read the history of the rocks and structures within the research area. Once the local history is known, the information can then be applied for future development or conservation of the area.

Summary

1. Geology is one of the earth sciences which is the overall study of the planet Earth and its position within the universe. Earth science also includes the sciences of astronomy (the universe), meteorology (weather) and oceanography (oceans).

2. Geology is the science of the Earth. It contains many sub-branches such as petrology (rocks), vulcanology (volcanoes), tectonics (deformations and movement of the surface), palaeontology (fossils), seismology (earthquakes) and many others.

3. Geologists use a wide range of tools, including satellites, GPS and electronic recording devices, sophisticated geophysical instruments and computers.

4. Field exploration by an individual or team with their instruments and an intuitive feel for variations in the land surface, sea and sky is still the most useful aid in developing this science.

5. Scientifically, thought and practice is constantly changing as the work of many people in the field and laboratory add to the overall understanding of the Earth. Vague ideas and questions which need to be tested (**hypotheses**) become more applicable (**theory**) and then established practices and knowledge (**law**) in time. However, as new information comes to hand, even established knowledge can be changed.

6. Measurement involves the accurate use of a variety of instruments, both analogue (visible scales) or digital (screen display) and users must be aware of the limitations and errors of instruments and record data in a significant and realistic way.

7. Measurement also involves the consideration of absolute and percentage errors (i.e. the absolute error of a measurement expressed as a percentage of its value) and the concepts of validity (is it true for the experimental aim) and reliability (can the experiment be reproduced).

8. Field work is usually carefully organised that the objectives of the research are achieved. This involves: clarification of objectives; a thorough review of previous research; and detailed preparation and planning of such matters as access to land, length of the research, equipment, communications, and personnel required, before the field work commences. Once the data has been obtained, it must be processed, analysed and communicated as a coherent conclusive work.

9. Part of the preparation involves the production of a base map which will be the framework for the data obtained during the research. From other maps, satellite and aerial photographs, the routes (or traverses) across the study area will be determined. These traverses may be open (going from one point to another, such as down streams) or closed involving several stages of legs of exploration which will return to the same starting point.

10. Safety in the field is paramount, especially if the research site is in a naturally hazardous area such as on a volcano or in a snow field. Some of the basic precautions would include: having a definite plan and giving to a responsible person; appropriate personal clothing and shelter with provision for unexpected changes in conditions; a good supply of food and water with emergency supplies; and good navigation equipment with a backup map in case the GPS fails.

11. One of the essential skills of the scientist is the power of observation. In the field, the earth scientist must look for subtle signs as well as the obvious. This includes: different outcrops of rock such as rounded tors, strata and volcanic peaks; changes in soil colour; changes in vegetation; and any sudden changes in the level of the land or the direction of rivers. Any linear feature often denotes something of interest.

12. Walking in the field may not be easy. Steep slopes, unstable surfaces, the vegetation cover and the climate all can make even a simple traverse very difficult and sometimes dangerous. Before the trip one can examine the slopes and conditions on maps and photographs and make good estimates of the degree of difficulty and the time that traverses will take. Even then, there will be unexpected situations.

13. Topographic or survey maps are the basic paper representation of the land surface. They can be stored and read electronically but the basic printed map is still the best choice. Apart from the usual information presented, such as heights (using contour lines), one should be familiar with using the scale, grid references and converting map bearings to magnetic bearings so a traverse can be taken using a compass.

14. Making a geological map of the research area is often one of the main tasks of the geologist or team. This could be a simple sketch map made in the field or a sophisticated, large scale geological map using satellite data and a thematic map printer to complement the field work. The map-maker must be knowledgeable of how the types of rocks, their orientation and disruption can be shown on a flat, plan view map which would communicate the nature of the mapped area truthfully and accurately. This may mean the proper use of the Brunton Compass, satellite and aerial photographs and data from drill cores taken in the area as well as precision in drawing.

15. Drawing a geological cross-section is an important skill which is useful in understanding the geological history of any area and being able to understand the bigger picture of the wider field including the large-scale tectonic forces which are shaping the area and the time frame in which the area formed.

Practical Tips

1. Basic field work is needed for any serious study. One must research the area to be studied with such material as notes from others who have studied the area, geological maps, satellite images and local knowledge;

2. A good geological hammer is essential. Cheap, student picks can be purchased at hardware stores and the like, but a serious geologist will purchase a good pick made of quality steel from a mining supplies store. This can be carried on a special belt and a good hand-lens on a neck lanyard is useful. If hunting for fossils, a broad brick chisel is better than the usual rock chisel. For navigation as well as rock measurements, a good compass is important. Usually the fluid-filled type with an in-built inclinometer used for measuring slopes and angles is preferred (e.g. a Brunton Compass). GPS and mobile phone Apps are useful but tend to be unreliable in some circumstances. Always carry a pocket notebook, pencil (not pens – ink runs when wet!) and any field maps and aerial photos that may be useful.

3. One must thoroughly prepare for the field trip and assume the worst in terrain and weather; take extra water, compressed and dry food, lightweight waterproof clothing, an emergency Space Blanket (aluminised plastic – pocket size), good boots, broad-brimmed hat (small and soft) and long pants and long sleeved clothing. All of this should only be able to fit into a small backpack with plenty of space for specimens.

4. Always make provision for unexpected changes in climate, the terrain and the functioning of electronic equipment. Additional water, food and some protective clothing and shelter can be easily carried even on a day trip e.g. basic equipment carried in a sturdy backpack (with good shoulder support and back frame) would include basic geology equipment, water (allow 2-4 litres/ day depending on availability and climate), food (concentrated food bars, nuts, dried fruit for 2 days minimum for a one-day trip), padded water-proof jacket (the type which rolls up into a small volume), rain shelter (the combine slicker-tent version), fire-making equipment (water-proof matches or a universal match of a flint/steel with fuel).

5. In cold weather, especially requiring an overnight camp in the field, carry additional water in an old-fashioned hot water bottle. It can be used at night in the sleeping bag as usual or in emergencies it can be drunk.

6. Sterilization pills (or potassium permanganate at last choice) are often useful if water is available but of dubious quality. Small sachets of coffee powder, sugar and powdered milk make this easier to drink.

7. Obtain an army map case or a wide sealable plastic bag to carry maps during wet or snowy weather. The army map cases often have a rigid, clear plastic sheet as a cover which can be written upon using china-graph (water-proof) pencils. A sturdy, water-proof cover for any electronic device is essential.

8. Have good communications such as a reliable mobile phone (check that it will be in range with local telco people) or radio (again check range). There is a the old adage that good communications will fail when you need it so always leave details of the trip and estimated time of return with friends and relatives and certainly with the local police or park ranger.

9. Always walk across country following a pre-planned route based on what will give the best observation based on prior research. One can take short trips off this route if something of interest is seen, but always return to the main route. The best route across country is usually along ridges and known roads, but some of the best outcrops occur along river courses, gullies and gorges. These are dangerous places as are highway and rail cuttings and old mine workings and quarries but they often yield the best results.

10. Don't be a stranger walking across the land. Visit locals and obtain their approval for walking across their land even if an exploration license allows you to do so. Most will not object and a considerable amount of good local knowledge and hospitality often is given. Avoid crossing the land of difficult land-owners or restless livestock; there probably are other routes that can be taken.

11. Plan field work so that it can be undertaken at a time giving the best conditions. If possible avoid times of extreme climate or the seasonal activity of wildlife (e.g. winter is a good time to explore snake country).

Multichoice Questions

1. A scientific statement or prediction of considerable doubt which requires more testing is called a(n):

    A. Observation
    B. Hypothesis
    C. Theory
    D. Law

2. In the field, a petrologist would be looking for:

    A. Gemstones
    B. Geological faults
    C. Rock types
    D. Fossils

3. A sedimentologist performing experiments on grain size, uses a 500 ml measuring cylinder to find the volume of some dune sand. The measuring cylinder is graduated (i.e. smallest unit marks on the glass) in units of 5 millilitres (ml). He knows that his measurement is approximately 250 ml.

Of the readings below, the most scientific expression of data and error is:-

    A. 250.0   +/- 0.5 ml
    B. 250.00 +/- 0.50 ml
    C. 250   +/- 5 ml
    D. 250.0   +/- 2.5 ml

4. The concept which is concerned that an experiment actually tests the truth of an hypothesis is called:

      A. Reliability
      B. Validity
      C. Accuracy
      D. Fallibility

5.   The <u>most reliable</u> navigational equipment which should be taken into the field is a:

      A. Current topographical map and Brunton Compass
      B. GPS unit
      C. GIS unit
      B. Geodometer and notebook

6. A closed traverse involves:

      A. Walking down a stream
      B. Walking along a ridge
      C. Being picked up at the end
      D. Returning to the start position

7. With the still air temperature being 5^0 Celsius and a wind speed of about 20 km/hour, the actual temperature would be about:

      A. 10^0 Celsius
      B. 5^0 Celsius
      C. -3^0 Celsius
      D. -5^0 Celsius

8. The estimated time that it would take to walk 8 km (from the map), through open country up a gentle slope from an altitude of 200 metres above sea level to 1100 m. a.s.l., would   be approximately:

      A. 2 hours
      B. 3 hours
      C. 4 hours
      D. 5 hours

9. This question refers to the following Topographic Map:

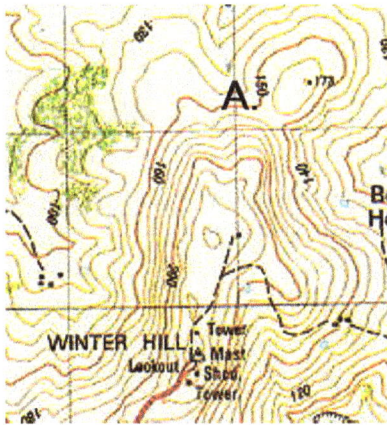

The geographical feature marked at A is most likely a:

      A. Conical Hill
      B. Spur
      C. Knoll
      D. Saddle

10. This question refers to the following diagram:

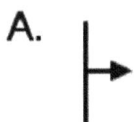

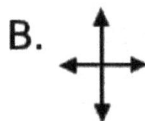

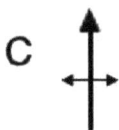

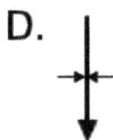

The symbol which represents a <u>plunging anticline</u> is:

A. A
B. B
C. C
D. D

Review and Discussion Questions

1. Geology can be an adventurous science. Is this true? Discuss.

2. Define each of the following terms:

    (a) seismology
    (b) vertebrate palaeontology
    (c) meteorology
    (d) geophysics

3. "The present is the key to the past" is a famous statement of which geological law? What does it mean in terms of application to modern earth science investigations?

4. Without knowing details of actual ages and time, Adam Sedgewick and others used names to represent rocks which had different relative ages (i.e. one was older than the next and so on). What was the source of these names? Are they still of use today? (Internet research may be needed).

5. "Things go wrong when they are least expected!" is a common expression. Discuss this in relationship to going into the field.

6. Why would a small-scale aerial photograph be an advantage over a topographical map of the same scale?

7. What basic equipment should be carried by a field geologist on a two day traverse in regard to:

(a) sampling geological data;
(b) basic living;
(c) safety;
(d) navigation

8. Define each of the following terms:

(a) contour interval;
(b) topography;
(c) symmetrical folds;
(d) batholith;
(e) vertical exaggeration.

9. A traverse along flat ground encounters dipping strata which is **not** dipping in the same direction as the traverse. The beds are all dipping at 30^0. What is the true dip if:

(a) the angle between the dip direction of the beds and the traverse is 20^0?

(b) the angle between the strike of the beds and the traverse is 60^0?

10. On another traverse, a bed of limestone dipping (true) at 40^0 outcrops for 120 metres along the traverse. What is its true thickness? How could that be confirmed?

11. Soils and vegetation can often indicate local geology. Discuss the role of local animals in indicating geology (an interesting question which might need some extra research).

12. What are the probable past environments indicated by these rocks:

    (a) greywacke;
    (b) volcanic breccia;
    (c) tillite;
    (d) mudstone with trilobite burrows.

13. State the Laws of:

    (a) Superposition;
    (b) Uniformitarianism.

Discuss why they are important in geological mapping.

14. Given a geological map for the first time, what are the things which a wise geologist would look for? What other actions might he/she take before seriously using the map?

15. This question refers to the following geological cross-section:

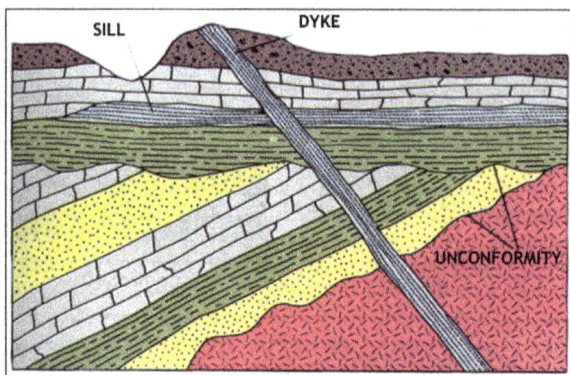

Give a sequential geological history of this area from oldest to youngest.

Answers to the Multichoice Questions

Q1. B Q2. C Q3. D Q4. B Q5. A Q6. D Q7. C Q8. C
Q9. D Q10. C

Reading List

Gohau, G. (1990). *A history of geology*. Revised and translated by Albert V. Carozzi and Marguerite Carozzi. New Brunswick: Rutgers University Press. ISBN 978-0-8135-1666-0.

Harbaugh, J.W. Geology. 2015. *Encyclopaedia Britannica*. http://www.britannica.com/science/geology

Iowa State University. (2015). Glossary of Geologic Terms. Dept. of Geological and Atmospheric Science. http://www.ge-at.iastate.edu/glossary-of-geologic-terms/

Janke, P.R. (1996). *Correlating Earth's History - Lecture Notes*. From a lecture given in May, 1996: *Learning from the Past*. Black Hills Museum of Natural History. http://www.wmnh.com/wmas0002.htm

Kent Educational, (2015). *Careers in Geology*. http://www.personal.kent.edu/~cschweit/Stark/whygeologymajor.htm

Lisle., R. J., Brabham, P. & Barnes, J. (2011). *Basic Geological Mapping* (5th Edition). Oxford. John Wiley & Sons. 235 p. ISBN: 978-0-470-68634-8

Lisle, R. J.(2004).*Geological Structures and Maps: A Practical Guide*. (3rd. Edition) Oxford, UK. Elsevier Butterworth-Heinemann., 106 p. ISBN 0 7506 5780 4

McLelland, C.V (Edit.).(2006). *The Nature of Science and the Scientific Method*. Geological Society of

America.
http://www.geosociety.org/educate/NatureScience.pdf

Mathez, A. (2000). *Earth: Inside and Out*. New York: New Press, American Museum of Natural History. Excerpts at:
http://www.amnh.org/education/resources/rfl/web/es saybooks/earth/

Ohio University. (2015). *Careers in Geology*. College of Arts and Sciences.
https://www.ohio.edu/cas/geology/careers/about.cfm

Pollard, D. D. & Fletcher, R. C. (2005). *Fundamentals of Structural Geology*. Cambridge U.K. Cambridge University Press 514 p. ISBN: 9780521839273

Roberts, J. L. (2013). *Introduction to Geological Maps and Structures*. Pergamon International Library of Science, Technology, Engineering and Social Studies.332 p. ISBN: 9781483140995.

Soller, D. R. (2004). *Introduction to Geologic Mapping*. USGS. (First published in the McGraw-Hill Yearbook of Science & Technology 2004, pp. 128-130.).
http://ncgmp.usgs.gov/geomaps/introgeo_mapping.html

Study.com.(2015). *Chapter 26: What is Geology?*
http://study.com/academy/lesson/what-is-geology-definition-history-facts-topics.html

Tearpock, D. J. & Bischke, R. E. (2002). *Applied Subsurface Geological Mapping with Structural Methods*

864 p. Upper Saddle River, NY. Prentice-Hall. ISBN: 0-13-859315-9.

Key Terms Index
(Page numbers in brackets)

accuracy (33) concerns how close a measurement is to the real value and depends upon the limitations of the instruments used to measure this quantity.

absolute error (38) is the total addition of all the individual errors of each measurement when measurements are added or subtracted.

analogue instruments (34) all of the scale of measurement can be seen and the user must judge the position of the measurement on the scale.

angle of dip (8,90) is the angle that strata is tilted, measured to the horizontal.

angular unconformities (119) occur when horizontally parallel strata of sedimentary rock are deposited on tilted and eroded layers.

apparent dip (93) is the false dip taken on tilted beds when measured on a plane which is not in the dip direction. apparent dip is less than the true dip and can be converted to this using tables and the direction angle (angle between the true dip direction and the direction of the apparent dip).

astronomy (1) is the study of the night sky and of the universe.

anticlines (103) are upward folding beds.

aureole (66) is the area around an igneous intrusion in which contact metamorphism occurs in the country rock.

back-bearings (81) are the reverse bearing of that taken to a feature i.e. the bearing to the feature - 180^0.

batholiths (59) are very large igneous bodies many kilometres across which have been formed deep below the surface from the cooling of molten magma.

closed traverse (54) is an exploration route which returns to its starting point. Enclosing an area of interest.

contact metamorphism (59 & 66) changes to existing rocks due to their proximity to hot igneous bodies.

Continental Drift (26) a theory which states that the modern continents were once a larger landmass called Pangea which broke apart and moved across the Earth's surface.

contour interval (75) the increment of vertical height between contour lines.

contour lines (72,75) are lines joining places of equal height.

cross-polarised (11) when a glass microscope slide is sandwiched between two polaroid filters which are orientated at 90 degrees to each other the light is polarised twice.

digital instruments (35) only the reading of the measurement is shown on a screen as a series of digits.

dip direction (90) is the geographical direction towards which the tilt of the beds is facing.

direction angle (97) is the angle between the true dip direction and the apparent dip direction.

disconformities (117) are unconformities between parallel layers of sedimentary rocks with little change in the angle of the bedding.

dykes (59) are vertical or near vertical tabular igneous rock intrusions which cut across normal horizontal beds.

Earth science (1) is the overall study of the Earth and of its place within the universe.

eastings (74) are the vertical lines of the grid on a map.

economic geology (2) is the study of Earth materials of value such as ores (economic minerals), valuable rock material and soils and the economic principles which govern their extraction and exploitation.

engineering geology (2) is the use of geological principles in construction.

environmental error (36) is an error of measurements due to adverse effects of the locality, its weather, terrain and other conditions.

Expanding Earth (27) hypothesis was used by S. Warren Carey to explain the spreading of the sea-floor without the need for subduction zones where oceanic plates are pushed below other plates. This hypothesis has not gained much acceptance.

fault scarp (108) is a cliff face formed by the movement of a fault up or down.

fence diagram (88) is a simulated three-dimensional sub-surface diagram of strata constructed by linking together the same rock beds and structures from different and separated drill cores.

Geographical Information System - GIS (45) hand-held electronic device for entering data in the field.

geochemistry (2) is the study of the materials of the Earth from the point of view of their atomic nature and interactions.

geodometers (51) are surveying instruments for very accurate measurement of distance. They often use a reflected LASER beam to calculate the distance from the time of reflection of the beam.

geological history (127) is a detailed sequence of rock formation and geological events from oldest to youngest event.

geology (1) the study of the surface and interior of the Earth

geophysics (2) is the study of the forces in the Earth.

Global Positioning System - GPS (45) involves a hand-held navigation device which uses triangulation from several satellites. These may have some limitations in remote areas and in extremes of weather.

gossan (60) is an iron-containing secondary deposit, largely consisting of oxides and typically yellowish or reddish, occurring above a deposit of a metallic ore.

gradient (69) is the ratio of the vertical height achieved on the slope to the distance covered horizontally on the map.

Grid North - GN (76) which shows direction of the vertical, numbered grid lines on the map.

Grid Magnetic Angle GMA (76) is the angle between Grid North and Magnetic North.

grid reference (74) is a 6 or an 8-digit figure which denotes a specific place on a topographical map using the grid squares marked upon it. the first 3 (or 4) numbers are taken from the horizontal line of grid numbers (eastings) and the last set of numbers from the vertical set of grid numbers (northings).

grid squares (74) are defined by the grid lines drawn on a map and are used for reference.

heave (110) is the horizontal displacement of a fault.

hiatus (116) is a gap in time or space (see unconformities).

historical geology (2) is that branch of geology which deals with detailing past events usually in sequential order.

Hubble's Law (27) states that the red shifts in the spectra of distant galaxies (and hence their speeds of recession) are proportional to their distance.

hypothesis (3) is an initial idea or question in science that needs to be tested by observation, measurement and experiment. Today, considerable computer modelling is used in an initial phase of this testing.

igneous rocks (61) are formed when molten rock material (magma if deep below the surface, lava if on the surface) cools.

instrument error (36) is an error of measurement due to the limitations in the construction of the measuring instrument.

karst topography (65) is the name given to the surface of limestone country which is riddled with open cave entrances and gorges.

LASER (44) is an acronym for Light Amplification (by) Stimulation (of) Emission Radiation.

law (3) a law is established, when over a long period of time, a theory has become reliable and works (within established limitations of error) every time it is applied. This does not mean that laws may not be challenged by further experimentation and new theories can be developed if a law is proven to be consistently wrong in new situations.

Magnetic North MN (76) points towards the north magnetic pole and is the direction that a compass needle will point if placed on the map.

magnetic declination or **variation (76)** is the difference in angle between True North and Magnetic North.

map legend (74) uses symbols to indicate some of the features on the ground.

marker beds (108) are recognisable beds seen on either side of a fault in cross-section which allows measurement of the vertical (throw) and horizontal (heave) movements of a fault.

metamorphic aureole (66,112) is the volume of change or metamorphism of rock around an intruding igneous body.

metamorphic rocks (66) are formed from existing rocks which have been changed by heat (contact metamorphism) or pressure with heat (regional metamorphism).

meteorology (1) is the study of the Earth's atmosphere and the weather patterns in it.

mineralogy (2) is the study of minerals and crystals.

Mohorovičić Discontinuity (25) is the boundary between the Earth's crust and mantle.

necks (59,63) are relatively small, circular igneous intrusions which are the remains of the central core or pipe of a volcano.

Neptunist Theory (4) named after the roman god of the sea, this theory suggested that rocks were formed as the waters of the great biblical flood evaporated, just as salt is left when seawater evaporates.

nonconformities (119) occur between sedimentary rocks and metamorphic or igneous rocks on which the new sediments have been deposited after the period of erosion.

northings (74) are the horizontal lines of the grid on a map.

north points (76) are the several indications of north e.g. True North, Grid North etc.

oceanography (1) the study of the sea floor and its features as well as the nature of the water above it.

open traverse (52) is an exploration route with a different start and finish.

operator errors (36) are due to the human condition and the observing ability of the person taking the measurement.

palaeontolgy (2) the study of ancient lifeforms (fossils) and their environments

paraconformities (118) are a type of disconformity in which the separation is a simple bedding plane with no obvious buried erosional surface.

Personal Data Assistant - PDA (45) is a portable personal computer for adding and processing data in the field. it also may allow for uploading data via satellite link.

petrology (2) the study of rocks e.g. igneous petrology (rocks from molten material), sedimentary petrology (rocks formed from sediment), and metamorphic petrology (rocks which have been changed from other rocks by heat and/or pressure).

plunge (107) is the angle of tilt of folded beds at ninety degrees to the angles of the folds.

plunging fold (103) is a folded bed which has also been tilted at right-angles to the fold.

Plutonist Theory (4) named after the roman god of the underworld this theory suggested that rocks were formed from molten rock material which came from deep below the Earth's surface and then cooled.

random errors (36) are the limitations due to the conditions prevailing during measurement such as environmental conditions.

regional metamorphism (67) is the change in rocks due to extreme pressure and/or heat.

relative error (38) or percentage error, is the addition of each error of a measurement (expressed as a percentage of the measurement) when values with different units are multiplied or divided.

reliability (42) of an experiment or measurement is its ability to be repeated and show some consistency in its data and calculations.

romer (75) is a reference card or scale marked on a mapping protractor which shows various scales for maps. it is used as a reference rule to subdivide grid squares on a map.

root mean square (38) is a statistical computation which uses only the magnitude or numerical value of data, not including their negative or positive signs. statistically it is the square root of all of the averages squared.

scale (74) on a map gives an indication of the distances on the ground.

scientific notation (31) is the system using powers of ten and their multipliers to describe very small and very large numbers.

Sea Floor Spreading (27) is a theory suggesting that the floor of the sea is formed at mid-ocean ridges by upwelling magma and that the convection in the Earth's mantle was the driving force behind this process.

sedimentary rocks (64) are formed from eroded materials later deposited (under water, or by ice and wind) and then compacted and cemented -usually in layers called strata.

seismology (2) is the study of Earthquakes and the effects of tectonics.

significant figures (29,40) in any measured value, these are the digit which convey the true meaning of that values and depend upon the limitations of the instrument(s) used in giving that value.

sills (59) are horizontal igneous rock intrusions which are between horizontal strata.

SI Units (29) are the standard units used in scientific measure and comes from the french *le système international d'unitésand* established in 1960.

smartphone apps (96) are compuer application or .exe files which can be downloaded to smartphones, computer tablets and PCs. There are many navigational, data and geological apps available.

stratigraphic column (88) is a representation of a cross-section through part of the Earth's crust, usually drawn from drill-core data.

stratigraphy (2) is the study of layers (strata) of rocks.

strike (8,91) is the geographical direction of the intersection of the beds with the surface (i.e. which way the outcrop of the beds appear to run along the surface of the land).

Superposition - Law of (127) states that in a sedimentary sequence, the youngest rocks overly the oldest rocks.

synclines (103) are downward folded beds.

systematic errors (36) are those due to the method of taking the measurement and may include the calibration (setting up the instrument scale etc.) of the instrument and its condition (e.g. battery charge, cleanliness etc.). These usually can be avoided or removed.

tectonics (2) is the study of deformation and movement of the surface layers giving   fractures and folding.

thematic mapper (72) is a device for gathering land-surface and often sub-surface data and is carried on board several landsat resources satellites. The data can be received on erath and maps can be printed directly.

throw (108,110) is the vertical displacement of a fault.

theory (3) is formulated once the testing of a hypothesis is successful in many cases. There still is considerable doubt, but a current theory is useful in making predictions about Earth events.

Theory of Continental Drift (26) states that the modern continents were once a larger landmass called Pangea which broke apart and moved across the Earth's surface.

topographical maps (50) are a representation of the land surface and most of the features found on it. Heights above sea level (a.s.l) are usually shown by contour lines (lines joining places of equal heights) but some simpler variations have colour shading.

tors (59) are large rounded boulders found on the surface which have been eroded from a large igneous body such as a batholith below.

traverse (50) is the planned route that a geologist will take to gather data along the best possible line(s). They may be open simply going from a to b, or closed as a circuit, of several legs, returning to the original starting point.

triangulation (81) is a method of using at least three back-bearings to locate a position on a map.

True North TN (76), which points towards the Geographical North Pole

unconformity (116) is a discontinuous boundary between rock strata representing a period of lack of sedimentation and erosion.

Uniformitarianism - Principle of (22) the geological processes observed today are the same as occurred in the past.

validity (41) of any observation, experiment or gathering of data, is how fairly or honestly the results tests the hypothesis in question.

vertical exaggeration (84,101) is the stretching of the vertical height of any cross-section so as to give a readable view. It is the ratio of the horizontal scale of distance over the vertical scale.

Vulcanist Theory (22) stated that all crystalline rocks formed from lava from volcanoes. This is another term used for the plutonist theory.

vulcanology (2) is the study of volcanoes, their causes and related features.

wind chill factor (58) is the degree by which the temperature of the body is dropped due to contact with the wind.

This book is also available in electronic format which can be purchased at amazonz.com for Kindle or other electronic devices such as PCs and iPad using the free Kindle App. Books in the series **ADVENTURES IN EARTH SCIENCE** are available from Felix Publishing, Australia (info@felixpublishing.com) and include:

EXPLORATION SCIENCE
Field Geology &
Mapping

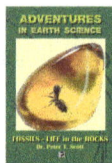
FOSSILS – LIFE in the ROCKS
Ancient lifeforms and their
environments

RICHES from the
EARTH
Minerals & Energy

A DANGEROUS PLANET
Volcanoes &
Earthquakes

CHANGING the
SURFACE
Erosion & Landscapes

THROUGH SEA and SKY
Oceanography &
Meteorology

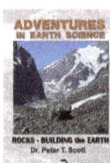
ROCKS – BUILDING
the EARTH
Rock types &
their composition

BEYOND PLANET
EARTH
An Introduction
to Astronomy

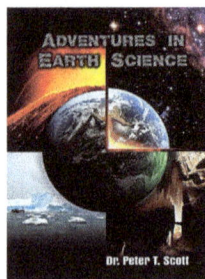
Now released in electronic and print (A4) format is the complete reference book and text book suitable for Senior Secondary Schools, Junior College and Universities – ADVENTURE IN EARTH SCIENCE. Over 800 pages, 1200 illustrations (mostly taken by the author in many exotic places on seven continents), and over 30 video links on skills and virtual excursions to many parts of the world.

www.ingramcontent.com/pod-product-compliance
Lightning Source LLC
Chambersburg PA
CBHW050726030426

42336CB00012B/1424